EMP Environment and System Hardness Design

By
Rabindra N. Ghose
Chairman of the Board
American Nucleonics Corporation

1st Edition

DON WHITE CONSULTANTS, INC.
State Route 625
P.O. Box D
Gainesville, Virginia 22065 USA
Telephone: (703) 347-0030
Telex: 89-9165 DWCI GAIV.

Library of Congress Catalog Card Number 83-51067
ISBN 0-932263-16-X

Printed in the United States of America

About the Author

Rabindra N. Ghose is founder of American Nucleonics Corporation. He was formerly associated with Radio Corporation of America, Ramo-Woolridge Corporation, Space Technology Laboratories, Space General Corporation, and University of Southern California. He holds a BEE (Summa Cum Laude) degree from Jadavpur University (Calcutta), a post graduate degree in Communication Engineering from the Indian Institute of Science, MS in Electrical Engineering from University of Washington, and a MA in Mathematics, PhD and post-doctorate EE degree from University of Illinois. He also holds an LLB degree and is a member of the California State Bar and a registered patent attorney.

Dr. Ghose is a Fellow of the Institute of Electrical and Electronics Engineers, Institute of Electrical Engineers (London), American Physical Society, American Association for Advancement of Science and Institute of Physics (London). He is a chartered Engineer in the British Commonwealth, and a Registered Professional Engineer in Electrical and Nuclear Engineering in California.

Dr. Rabindra N. Ghose

To my wife and son—

Dorothy and Geoffrey

Foreword

It is a pleasure, on behalf of Don White Consultants, Inc., (DWCI) to release this handbook on *EMP Environment and System Hardness Design*. This is another of our many published handbooks on Electromagnetic Compatibility which will provide a major contribution to the literature for years to come.

This handbook, *EMP Environment and System Hardness Design,* fills an existing void. It was prepared for engineers and scientists, military and civilian organizations involved in physical vulnerability assessments and for system designers, managers and users of equipment who may be concerned about nuclear effects on their systems and how to *harden* against such effects. It is written in terms of the advanced state-of-the-art and carefully illustrated so it will have broad applications in this important and dynamic substantive area. The author, Rabindra N. Ghose, invites your comments. I also welcome correspondence from any of the many readers who may want to comment on any aspect of this book.

January 1984
Gainesville, Virginia USA

Donald R.J. White
Publisher

Handbooks Published by DWCI

(1) White, Donald R.J., *Electrical Filters—Synthesis, Design & Applications,* 1980.

(2) White, Donald R.J., Volume 1, *Electrical Noise and EMI Specifications,* 1971.

(3) White, Donald R.J., Volume 2, *Electromagnetic Interference Test Methods and Procedures,* 1980.

(4) White, Donald R.J., Volume 3, *Electromagnetic Interference Control Methods & Techniques,* 1973.

(5) White, Donald R.J., Volume 4, *Electromagnetic Interference Test Instrumentation Systems,* 1980.

(6) Duff, Dr. William G. and White, Donald R.J., Volume 5, Electromagnetic *Interference Prediction & Analysis Techniques,* 1972.

(7) Hill, James S. and White, Donald R.J., Volume 6, *Electromagnetic Interference Specifications, Standards & Regulations,* 1975.

(8) White, Donald R.J., *A Handbook on Electromagnetic Shielding Materials and Performance,* 1980.

(9) Duff, Dr. William G., *A Handbook on Mobile Communications,* 1980.

(10) White, Donald R.J., *EMI Control Methodology & Procedures,* 1982.

(11) White, Donald R.J., *EMI Control in the Design of Printed Circuit Boards and Backplanes,* 1982. (Also available in French.)

(12) Jansky, Donald M., *Spectrum Management Techniques,* 1977.

(13) Herman, John R., *Electromagnetic Ambients and Man-Made Noise,* 1979.

(14) Hart, William C. and Malone, Edgar W., *Lightning and Lightning Protection,* 1979.

(15) Kaiser, Dr. Bernhard E., *EMI Control in Aerospace Systems,* 1979.

(16) Feher, Dr. Kamilo, *Digital Modulation Techniques in an Interference Environment,* 1977.

(17) Gard, Michael F., *Electromagnetic Interference Control in Medical Electronics,* 1979.

(18) Carstensen, Russell V., *EMI Control in Boats and Ships,* 1979.

(19) Georgopoulos, Dr. Chris J., *Fiber Optics and Optical Isolators,* 1982.

Handbooks Published by DWCI

(20) Mardiguian, Michel, *How to Control Electrical Noise,* 1983.

(21) Denny, Hugh W., *Grounding for Control of EMI,* 1983.

(22) Ghose, Rabindra N., *EMP Environment and System Hardness Design,* 1983.

(23) Mardiguian, Michel, *Interference Control in Computers and Microprocessor-Based Equipment,* 1984.

(24) *EMC Technology 1982 Anthology*

(25) Mardiguian, Michel, *Electrostatic Discharge—Understand, Simulate and Fix ESD Problems*

(26) White, Donald R.J., *Shielding Design Methodology and Procedures*

Notice

All of the books listed above are available for purchase from Don White Consultants, Inc., State Route 625, P.O. Box D, Gainesville, Virginia 22065 USA. Telephone: (703) 347-0030; Telex: 89-9165 DWCI GAIV. For countries outside North America, contact Els de Groot, DWCI Holland, Kerkstraat 63-65, 2355 AH Hoogmade, Holland. Telephone (31) 1712 2526; Telex 844-30268.

Preface

Electromagnetic pulse resulting from a nuclear detonation, often referred to by its acronym EMP, is one of the most serious nuclear effects of concern to electronic and weapon systems, since it can adversely affect performances of such systems at distances hundreds of miles away from the source. Although the mechanism of the EMP generation has been known for some time, there has been a considerable progress in recent years to more fully understand the characteristics of the EMP and to assess its damaging potentials. From such an understanding evolved the concept of EMP hardness, which symbolizes the safeguard or protection of systems or equipment against EMP during and following a nuclear burst.

Technical activities in the EMP field for the past decades have been in four broad areas. The first area addressed the characterization of the EMP depending on the source and the environment such a source creates, for nuclear detonations at different altitudes above the earth's surface, including the ground burst. The second area involved the analyses of various coupling mechanisms for the EMP energy into electronic and weapon systems. Technical works in the third area addressed the problem of assessing the susceptibility or vulnerability of systems due to the EMP, and investigations of techniques and the methodology required to protect the system from various EMP threats. The fourth area encompassed the development of appropriate tests, sensors and facilities where qualitative and quantitative EMP effects on systems or components can be evaluated by physical simulation of the EMP. The literature containing the results of such technical activities, however, are scattered. Past attempts at compilation of the EMP information became too voluminous to be contained in one book. The practical engineer, the system manager or the user of a system, concerned with the EMP problem and plausible remedies, may be better served by a short concise book on the subject which not only provides an understanding of the nuclear EMP but also a step-by-step methodology to assure the system hardness against the EMP during the design and acquisition phase and during the system life-cycle. The object of this handbook is to attempt to fulfill such needs.

The nuclear detonation-altitude and the yield primarily determine the intensity and waveform of the EMP and therefore define the EMP threat against which the system has to be protected. Various probable sources of EMP with different detonation altitudes are discussed in Chapter 2. For given EMP characteristics, the susceptibility or vulnerability of an electronic or weapon system will depend on the mode of entry of the EMP energy into the system. Various coupling characteristics which determine such mode of entry are discussed in Chapter 3. Effects of the EMP on systems in terms of "upsets" or permanent damages are discussed in Chapters 4 and 5.

The knowledge of the characteristics of EMP, its coupling mode, and the interaction with the system suggests means which may harden a system against the EMP. Here the hardness refers to the invulnerability of the system against the threat. Various hardness concepts and remedies to assure hardness are discussed in Chapters 6 and 7. The methodology for the formulation of standards and specifications required to assure the system hardness during its acquisition phase and life-cycle, and the EMP hardness-assurance management concepts, are also discussed in these chapters. Various system tests, under appropriately simulated EMP environment, constitute another important aspect of the EMP hardness assurance. Test concepts, sensors, instrumentations and required test facilities are presented in Chapter 8.

This handbook is intended for system designers, managers, and users who may be concerned about the nuclear EMP effects on their systems and remedies for relief. The development of a physical understanding of the EMP problem and modes or means of avoiding the consequential adverse effects is attempted in this handbook. Complex mathematical analyses are therefore avoided, and only an elementary knowledge of mathematics for the reader is assumed.

Numerous technical notes and papers on the subject of EMP, notably many EMP Sensor and Simulation Notes and EMP Handbook for Missiles and Aircraft in Flight, published by U. S. Air Force Weapons Laboratory, and the lecture notes of Dr. A. Sankar on EMP, privately communicated to the author, have been freely consulted during the preparation of this book. Grateful acknowledgement is made to the authors of these sources. Sincere thanks are also due to Norma Faul, Barbara Micheletti, Diane Nissen, Elaine Regan, Beverly Wanamaker and Marian Winter, all of American Nucleonics Corporation, for their assistance in the preparation of the manuscript and Geoffrey Ghose for his help toward the preparation of some illustrations.

Westlake Village, California — Rabindra N. Ghose

January 1984

Table of Contents

EMP Environment and System Hardness Design

Illustrations & Tables

Chapter 3 Modes of Entry of EMP Energy into Systems

Chapter 6 Hardness Concepts and Techniques

Chapter 7 EMP Hardness Assurance Program

Chapter 8 EMP Testing and Evaluation

Abbreviations

CEP	Circular-Error Probability
cm	Centimeter
CW	Continuous Wave
E	Electric field intensity
ECM	Electronic Countermeasures
EHF	Extremely High Frequency
ELF	Extremely Low Frequency
EM	Electromagnetic
EMC	Electromagnetic Compatibility
EMI	Electromagnetic Interference
EMP	Electromagnetic Pulse
GHz	Gigahertz
H	Magnetic field intensity
HPD	Horizontally Polarized Dipole
Hz	Hertz
IEMP	Internal Electromagnetic Pulse
K	Kelvin scale or kilo
KHz	Kilohertz
km	Kilometer
kt	Kiloton
KV	Kilovolt
LF	Low Frequency
M	Mega
MeV	Million Electric Volt
MHz	Megahertz
MT	Megaton
RFI	Radio Frequency Interference
SGEMP	System Generated Electromagnetic Pulse
TEM	Transverse Electromagnetic
TEMPS	Transportable EMP Simulator
UHF	Ultra High Frequency
V/A	Volt/Ampere
VHF	Very High Frequency
VPD	Vertically Polarized Dipole

CHAPTER 1

Introduction and Historical Background

Nuclear Electromagnetic Pulse (EMP) is a high-intensity, short-duration electromagnetic field that originates from a nuclear detonation. Compton current resulting from the gamma radiation following a detonation is the principal source of EMP for high altitude nuclear bursts, whereas the interaction of the highly conducting expanding plasma with the earth's magnetic field constitutes the basic mechanism of EMP generation from an underground nuclear detonation. Although the energy content of EMP is not very large because of its short duration, very high voltages or currents can be induced in components of electronic or weapon systems when EMP energy couples into such systems. These voltages or currents can, in turn, cause temporary or permanent damages affecting system missions. From the characteristics of EMP and the knowledge of the system under consideration, one can estimate EMP vulnerability of such a system. Theory and means now exist to reduce EMP vulnerability of a system. Analytical tools and step-by-step methodologies developed during the 1960's can be used to obtain and maintain EMP hardness of a system for any given threat. To assure EMP hardness, suitable tests are needed. Many different EMP simulators and test procedures have been developed over the years for such a purpose.

1.1 Nuclear EMP

The Nuclear Electromagnetic Pulse (EMP) is an intense time-varying electromagnetic field which originates from a nuclear detonation. The concept of the pulse in the nomenclature of EMP is perhaps due to its very short duration at any receiving point, the duration of the field being seldom more than a few microseconds and much of the EMP energy being concentrated within a small fraction of a microsecond. Although an EMP may be caused by non-nuclear explosions as well, the present usage of the term EMP is such that it refers to EMP of nuclear origin exclusively.

As an electromagnetic field, the EMP satisfies the physical laws governing the radiation and propagation of electromagnetic waves. The recognition of EMP as an intense electromagnetic field at the outset is useful since a vast reservoir of knowledge and informtion on the radiated Electromagnetic Interference (EMI), which is also an electromagnetic field, can be used directly or by valid extrapolations, to determine the impact of EMP on electronic and weapon systems. It should be recognized, however, that, because of the severe intensity of the EMP field, the difference in practical problems due to the EMP and radiated EMI may be very profound.

To determine system vulnerabilities due to an EMP or to design system hardness against the EMP threat, one needs to ascertain the characteristics of the EMP and the energy coupling to various components and subsystems for such specific EMP characteristics. One must also determine quantitative effects of the EMP-coupled energy on the degradation of system or mission performance. Characteristics of the EMP are determined by the nature of the electromagnetic source, source mechanisms, and the radiation characteristics of the field from the point of detonation to the receiver or the collector of EMP energy.

Explosions of even conventional high explosives can produce electromagnetic fields by interaction with the earth's magnetic field. The generation of EMP from a nuclear detonation is, therefore, to be expected in principle. Indeed, nuclear explosions of all types - from underground to high altitude - are accompanied by an EMP, although the predominant mechanism responsible for the generation of EMP is different, and the intensity and duration of the pulse vary with the location of the burst altitude. Compton current resulting from the gamma radiation following a detonation, for example, is the principal mechanism of EMP generation for the high altitude burst; whereas the interaction of the infinitely-conducting expanding plasma with the earth's magnetic field is believed to be the principal mechanism for the generation of EMP from an underground nuclear detonation. Quantitative aspects of various mechanisms of EMP generation are further discussed in Chapter 2.

1.2 EMP and its Impact on Electronic Systems

The characteristics of EMP which are of concern for the protection of electronic and weapon systems depend on the radiation characteristics, in addition to the source mechanism, and the yield and location of the burst of the nuclear device. The EMP radiation is somewhat similar to the familiar radio, although there are some important differences. For example, conventional radio transmitters are designed to generate electromagnetic waves of a particular frequency. The field of EMP has a wide range of frequencies and amplitudes. Also, the strength of the EMP can be millions of times greater than that of an ordinary radio signal. In each case, however, the energy of the electromagnetic waves is collected by a suitable antenna and is transferred to adjacent equipment. Although the EMP energy is not that significant, it causes a very strong current in components of a system, because of its very short duration. A typical short-duration current pulse induced by EMP from a high-altitude nuclear explosion in a long overhead power line is shown in Figure 1.1.

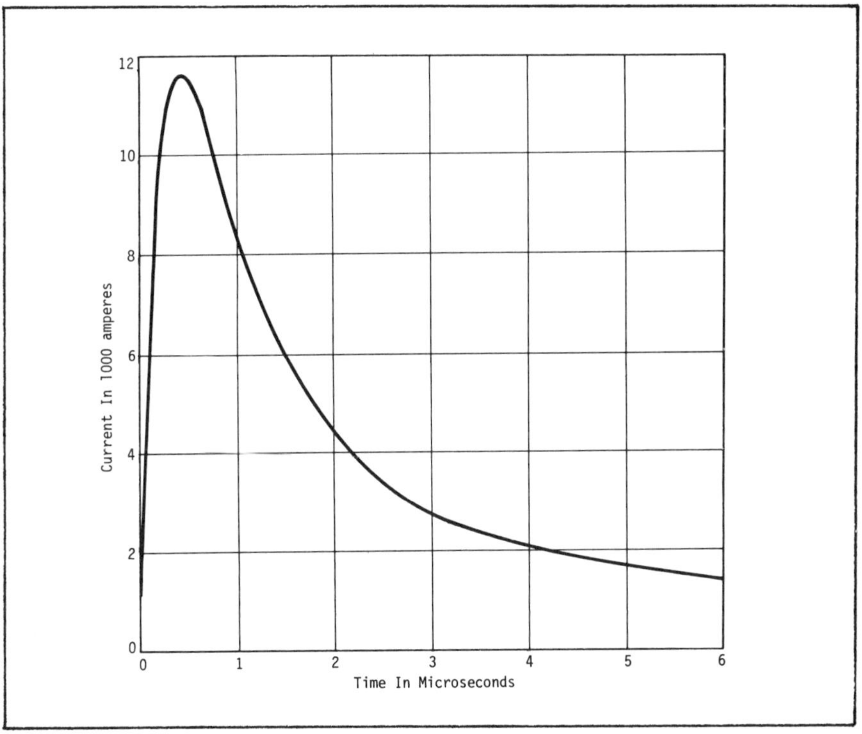

Figure 1.1 - Typical magnitude and duration of current induced by high-altitude-burst EMP in a long overhead power line. The magnitude will depend on yield, distance of burst, length of line exposed to EMP, etc.

The intensity of the EMP field is so large that often it does not need an antenna to collect the damaging energy. For example, it is not unusual for the EMP energy to reach a circuit or equipment inside a shield or enclosure by a direct penetration through the shield or through cables leading to the equipment or through doors, windows and seams of shield structure. Since the early 1960's, many theories, along with their experimental verifications, have been advanced to assess quantitatively various modes of entry of EMP energy into circuits and components. The exact computation of the field or voltages or currents induced in a circuit or component of a system, by almost any mode of entry, is difficult, even when the incident, time-varying EMP field is exactly defined. This is often because of the complexities of the formulation of the electromagnetic boundary value problems for objects with irregular shapes and sizes and the simultaneous presence of multiple modes of entry of EMP energy. For example, one of the more simple problems, involving the direct penetration of EMP field through a shield or an enclosure for a given incident field at the outer surface of the shield, can be solved exactly for only a few ideal geometries, such as an infinite circular cylinder, sphere, etc., most of which are not encountered in real-life systems. Similar difficulties arise with the analytical treatment of determining EMP-induced voltages and currents at the output of an arbitrary, but conventional, antenna, or in cables of arbitrary lengths and with arbitrary, but commonly occurring, terminations. Theories developed during the 1960's, however, do provide an insight into the damaging potential of the EMP by various modes of entry and suggest what measures are important during the design of a system to assure EMP hardness. The subject of various modes of entry of EMP energy into systems is dealt with in more detail in Chapter 3.

1.3 Vulnerability and System Hardness

Very often, the exact, or nearly exact, magnitudes of voltages, currents or fields induced by the EMP in a circuit or equipment is not of concern to the user and designer of the system, since their ultimate objective is to determine the vulnerability impact of the EMP in their systems and possible remedies to eliminate, deter or circumvent such an impact when and as needed. In this handbook, the vulnerability evaluation problem due to a nuclear EMP is divided into two parts: susceptibility and vulnerability. Susceptibilities refer to the direct cause-effect relation of the EMP field, such as a transient circuit upset or latch-up as the EMP strikes, or permanent damages in the form of burning a transistor or similar components in a circuit or a system. Some such susceptibilities may not be fatal, or even significant, to the mission or functional requirements of the system. For example, a warning receiver may complete its mission before it is destroyed partially or completely by an EMP without affecting the mission. Similarly, a temporary circuit upset or latch-up due to EMP, where the normality is restored within a short period, may not affect the ultimate mission or performance requirements of the system. Therefore, the vulnerability impact of the EMP under such circumstances may not exist or may at least be insignificant. Susceptibility considerations and evaluation methodology as discussed in Chapter 4 are, however, important, since they constitute the starting point for the vulnerability assessment.

The EMP vulnerability of a system as referred to in this handbook relates to the probability of failure to achieve a certain level of mission success. The mission of a defensive weapon system against a missile, for example, is the probability of detecting the on-coming missile on time, the probability of launch of the defending missile, the probability of its successful guidance for an effective strike, etc. If the EMP susceptibilities of detection, launch and guidance systems are such that their combined effect fails to prevent the attacking missile from reaching the target, the defensive missile system becomes vulnerable. Since the elements constituting the vulnerability of a system are probabilistic in nature, the evaluation of system vulnerability may also be made in terms of probability. The formulation of variables that affect vulnerability, a methodology for vulnerability prediction, and the survivability of a system ensemble are important elements of the vulnerability evaluation process and are further discussed in Chapter 5.

Once the free-field EMP characteristics are known, and the equipment or circuit susceptibilities and consequential vulnerability of a system for a given mission are determined, one can address the problem of hardness assurance which will safeguard or protect the system against the EMP, both during and following a nuclear burst. EMP hardness assurance problems can be grouped under three distinct categories, depending on the current status of the system which needs EMP protection. The first category or problem arises from the retrofit requirements for existing systems to assure EMP vulnerability. A preliminary examination of the system and an expected worst-case EMP environment may suggest, for example, fixes of certain components and subsystems. Depending on the degree of EMP susceptibility of such components or subsystems, one may use shieldings, filters, voltage limiters and circumventing techniques. Useful means available for such purposes are outlined in Chapter 6.

The second category of EMP hardness assurance problems is encountered during the design and implementation of new systems not in existence now or not even adequately defined at present. For such systems, some basic EMP hardness means or remedies can be integrated into the system design instead of employing them as after-thoughts. Good design practices, including appropriate grounding, which improves system integrity regardless of the presence or absence of EMP threats, can be incorporated into the design at the outset.

The third category of problems relating to EMP hardness emerges from the requirements of EMP hardness maintenance during the entire life-cycle of an electronic or weapon system, particularly when the system must undergo changes to accommodate new technological innovations or to remedy obsolescence. The concept of the EMP vulnerability control plan and its administration, along with other technical remedies to assure EMP hardness, evolved as solutions to this category of problems. As in the case of electromagnetic interference control, the formulation of standards and specifications becomes necessary during the system design and initial acquisition. Guidelines to except parts of such specifications for cost-effective considerations, and provisions to accommodate changes in a system during its life-cycle without compromising the EMP hardness, need to be incorporated in such standards and specifications. Methodologies for the formulation of standards and specifications are further discussed in Chapter 7.

1.4 Hardness Verification

To achieve the desired level of confidence that a system is designed and implemented properly for the EMP hardness, some verification testing is essential. Since it is unrealistic to verify the EMP hardness of a system in a true nuclear EMP environment, the simulation of the EMP environment for test purposes becomes a necessity. Again, since most simulation environments create only a part of the true environment, test objectives and meaningfulness of the test data, usable directly or by extrapolation, to interpret the degree of hardness, need to be established. In addition, because of the intense field of the EMP for an extremely short duration, appropriate sensors and instrumentations and measurement technology become necessary for EMP tests. Significant developments on EMP test concepts, simulation technology and sensors have been made in recent years. They are briefly described in Chapter 8.

Although the possibility of the EMP has been predicted even in early nuclear detonation (Trinity test, 1945), much of the history of understanding the EMP phenomena and technologies that evolved around the EMP spans a relatively short period. Major EMP developments in understanding of EMP phenomenology, environment, simulation, and analysis to determine the EMP-coupled energy into systems, were made during the early 1960's. By the late 1960's, plans were already underway to implement the hardness assurance concepts in a few systems. Today, users of systems, particularly those which are required to survive functionally during or following a nuclear detonation, are convinced of the menacing effects of the EMP and the necessity of EMP hardness of a system. Adequate understanding of EMP effects, analytical tools and test data, although scattered, are now available. A survey and organization of this information are attempted in the following chapters.

1.5 References

1. Longmire, C. L., *On the Electromagnetic Pulse Produced by Nuclear Explosion*, I.E.E.E. Trans. AP-26, 1978.

2. Brown, G. L., Ghose, R. N., Green, J. X., Putzer, E. J., Schwartz, K., and Wouters, L. F., *Electromagnetic Pulse Energy Studies*, American Nucleonics Corporation, Glendale, California, February 1967.

3. Vance, E. F., *EMP Hardening of Systems*, S.R.I. International, California, Fourth Symposium and Technical Exhibition on Electromagnetic Compatibility, Zurich, March 1981.

CHAPTER 2

Sources of EMP and EMP Environment

Prompt gamma rays which follow the nuclear detonation are the principal source of EMP. For the surface, air, and high-altitude nuclear bursts, the gamma radiation leads to bursts of electron flow, the electrons being generated by the Compton effect, photoelectric effect and by the "pair-production" effect. Of these effects, Compton electrons, following Compton effect, constitute the principal driving mechanism for EMP. In the absence of atmospheric nonhomogeneities, the electron flow from the detonation will be radially symmetric. Such an electron distribution creates a local radial field, but not the radiated EMP, since no magnetic field can exist for such a case. In real life situations, however, there is always some asymmetry due to the presence of the earth's surface, vertical gradients in atmospheric density, etc. As a result, the electron flow will tend to be more vertical than horizontal, creating an effective vertical electric dipole. The EMP may be considered to be radiated from this dipole.

The internal EMP, the system-generated EMP, and the underground EMP generation mechanism, however, are substantially different from those which result from the surface and above-surface bursts, although the gamma radiation is still the principal source for such EMP's. Characteristics of EMP depend on the yield of the nuclear weapon and the altitude of bursts for the surface and above-surface bursts. The internal EMP and the system-generated EMP also depend on the yield, but they also depend on the construction of the system enclosures, and walls, asymmetry of the enclosures, nature of shieldings, etc. The underground EMP, where the products of the detonation are contained, results from the magnetohydrodynamic effect and from currents induced by gamma radiation on the metallic casing of the test hole.

2.1 Mechanism for the EMP Field

The principal driving mechanism of EMP generation is Compton electrons, generated by nuclear-detonation-induced gamma rays interacting with matter, which is usually the air in the atmosphere. The Compton electrons, in turn, may generate other free electrons by electron collisions with other atoms. To a lesser extent, the photoelectrons generated by the weapon-generated X-rays may also contribute toward EMP. An EMP may also be generated by the magnetohydrodynamic effect where a highly conducting plasma, resulting from a nuclear detonation, momentarily distorts or displaces the earth's magnetic field. This magnetic field then relaxes back to its original position after the plasma cools. The rate of change of this magnetic field becomes a source of EMP. The mode of EMP generation by the earth's magnetic field displacement is considered a secondary mechanism, the primary mechanism being the Compton electrons generated by the prompt gammas from a nuclear burst.

The fission of uranium, or plutonium or the fusion of isotopes of hydrogen in a nuclear weapon leads to the liberation of a large amount of energy in a very short period of time, and within a limited quantity of matter. Regardless of the height of the burst, approximately 85 percent of the explosive energy of a nuclear fission weapon produces air blast, thermal radiation and heat. The remaining 15 percent of the energy is released as various nuclear radiations. Of this, 5 percent constitutes the initial nuclear radiation lasting for a minute or so. The remaining 10 percent constitutes the residual or delayed nuclear radiation which is emitted over a period of time. A part of the initial nuclear radiation becomes the source of EMP.

The initial nuclear radiation consists mainly of gamma rays which are electromagnetic radiations of high energy. The calculated gamma-ray energy output per kiloton energy yield from a hypothetical nuclear explosion is shown in Figure 2.1. Since the instantaneous or prompt gamma rays emitted in the nuclear reactions, and those produced by neutron interactions with weapon residues and the surrounding medium, are responsible for EMP generation, it is not hard to imagine the impressive available energy for EMP source, particularly for large nuclear bursts with yields in megatons. Only a small fraction of this available energy, however, appears in the form of the radiated EMP. For example, only about one hundredth of the prompt gamma-ray energy for the high-altitude burst and 10^{-7} of the prompt gamma ray energy for the surface burst appears in the radiated EMP. Quantitatively, the radiated EMP energy from a high altitude burst of a megaton yield is about 10^{11} joules.

2.1.1 Types of Gamma-Ray-Matter Interaction

There are three important types of interactions of gamma rays with matter as a result of which photons are scattered or absorbed. The first of these is called the "Compton effect". In this interaction, the primary gamma-ray photon collides with an electron and the energy of the photon is transferred to the elctron. A secondary photon, with less energy, then moves off in a new direction at an angle to the direction of motion of the primary photon. This process is shown schematically in Figure 2.2. The Compton interaction, therefore, results in a change of direction or scattering of the gamma-ray photon and a degradation of energy. The electron, after colliding with primary photon, recoils in such a manner as to conserve energy and momentum. Such an electron is called the Compton electron.

The second type of interaction of gamma rays and matter is photoelectric effect. By this process a photon with an energy somewhat greater than the binding energy of an electron in an atom, transfers all energy to the electron, which is ejected from the atom as a consequence.

The third type of interaction of the gamma radiation with matter is called "pair-production". When a gamma-ray photon with energy in excess of 1.02 MeV (million electric Volt) passes near the nucleus of an atom, the photon may be converted into matter with the formation of a pair of a positive and negative electrons. As with the photoelectric effect, the pair-production

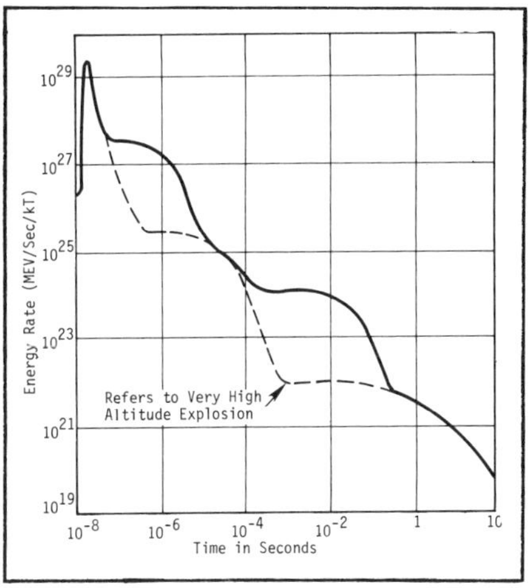

Figure 2.1 - Calculated Gamma-ray Energy Output Per Kiloton Energy Yield From a Hypothetical Nuclear Explosion As a Function Of Time.

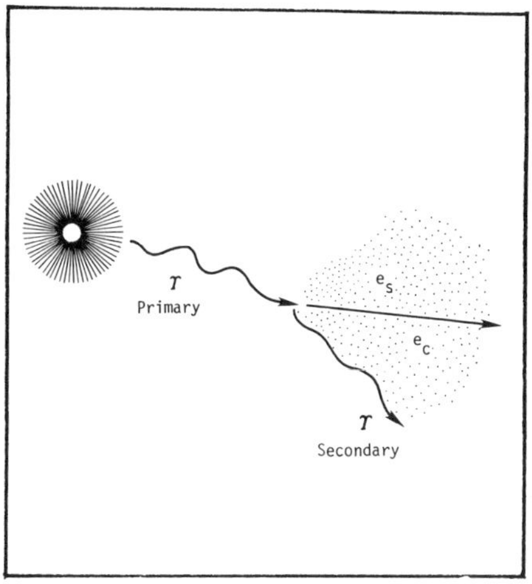

Figure 2.2 - Compton Effect, e_c And e_s Denote Compton Electron And Secondary Electrons Respectively.

results in the disappearance of the initial gamma-ray photon. However, the positive electron soon interacts with a negative electron forming two photons of lower energy than the original one. Of the three types of gamma-ray-matter interaction, the Compton effect is the most significant insofar as EMP generation is concerned.

2.1.2 Deposition Region and EMP

The EMP is considered to be generated at the "deposition region", where the gamma radiation from the weapon deposits its energy and produces an ionized region surrounding the burst location, mainly by the Compton effect. Figure 2.3 shows the extent of the deposition region for surface or near-surface bursts. Here, the gama rays interact with air and produce a radial-electric field. The deposition region as shown in the figure, is hemispherical, due to the asymmetry introduced by the almost infinitely absorbing and infinitely conducting ground plane at the earth's surface. This asymmetry causes the vector sum of all electron currents along the vertical direction and generates, thereby, a ϕ-component of the magnetic field. For the air burst, there will be a more radial symmetry of the electron distribution than is experienced in a surface burst. The deposition region for typical air bursts is shown in Figure 2.4.

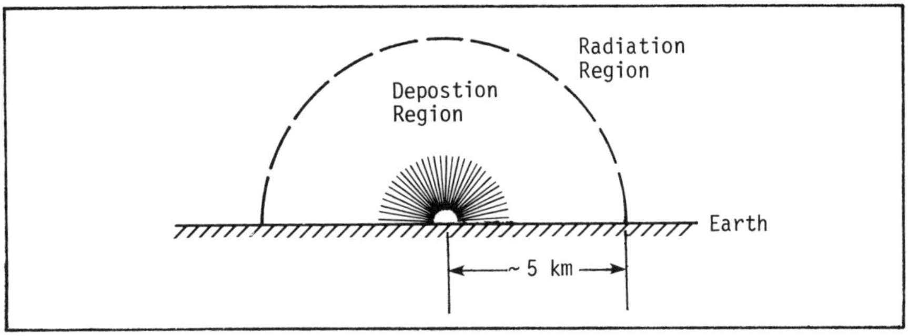

Firgure 2.3 - Deposition Region, Source Of EMP For A Ground Burst.

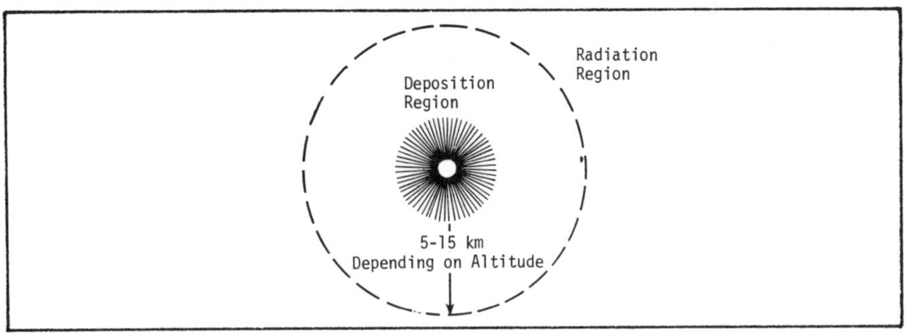

Figure 2.4 - Deposition Region For An Air Burst.

The deposition region for the high-altitude burst varies greatly as a function of the height of burst and the nuclear weapon yield. Typical extents of the deposition region for nuclear weapons of 1 to 10 megaton yields and for various heights of bursts are shown in Figures 2.5 and 2.6. As one may note, the deposition region in a high-altitude burst may be many times larger than that expected for the surface burst and even for the air burst.

Outside the deposition region, EMP appears as a radiated electromagnetic field. In other words, EMP waveforms have a 1/R dependence, R being the distance between the detonation region and the point of observation of EMP.

Recalling that EMP is an electromagnetic field, there must be an equivalent electric or magnetic current source, or both, for the generation of

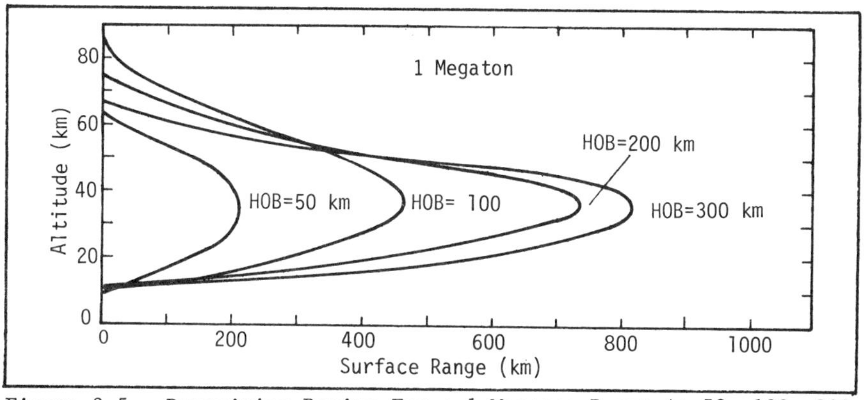

Figure 2.5 - Deposition Region For a 1 Megaton Burst At 50, 100, 200 And 300 km Height Of Bursts.

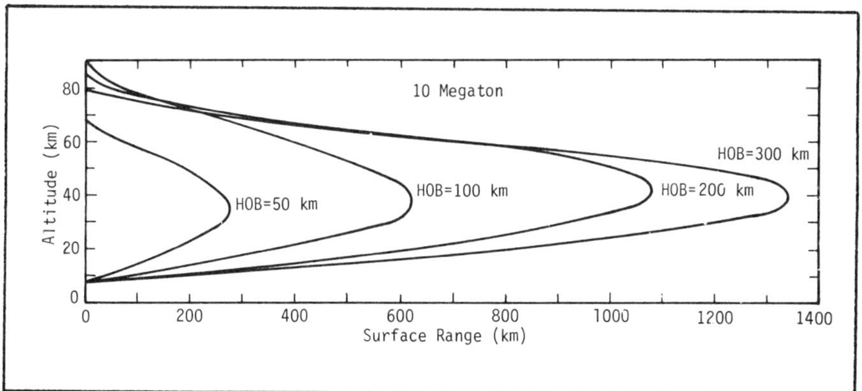

Figure 2.6 - Deposition Region For A 10 Megaton Burst At 50, 100, 200 And 300 km Height Of Bursts.

such a field. The moving electrons resulting from the initial gamma radiation, as discussed earlier, become this current source. Once the characteristics of such a source is determined, the electromagnetic field resulting therefrom can be evaluated from Maxwell's equations and the appropriate boundary conditions.

It should be noted, however, that not all electric current distributions create an EMP. For example, if one considers an idealized case of a nuclear detonation in a homogeneous atmosphere that extends to infinity in all directions as shown schematically in Figure 2.7, the radial electron flow in the deposition region will be spherically symmetric. It is the basic property of the electromagnetic field that a spherically symmetric current distribution does not create a magnetic field. Such a current distribution, therefore, cannot create a radiated EMP.

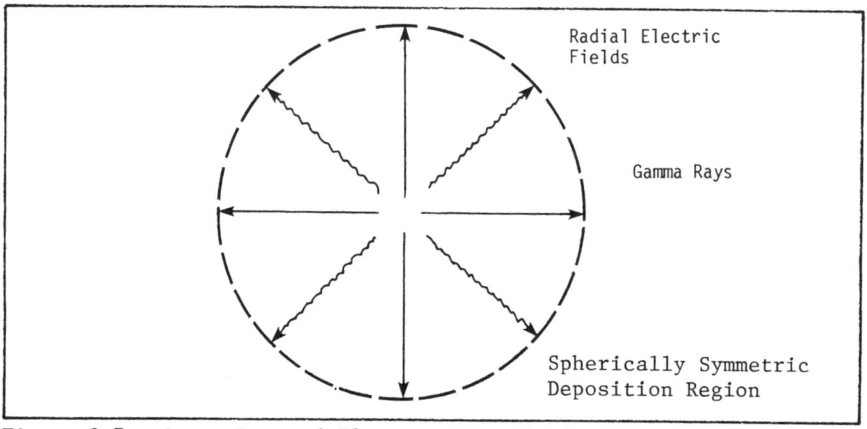

Figure 2.7 - Gamma Ray and Electron Distributions From a Hypothetical Nuclear Detonation In A Homogeneous Atmosphere.

In a more realistic environment encountered by a nuclear detonation, there will not be an absolute or even a near absolute symmetry of the electron flow. Asymmetry in such cases may result from atmospheric nonhomogeneities such as the earth's surface, gradients of water vapor and atmospheric density, the earth's magnetic field, and the configuration of the nuclear weapon and vehicle carrying the weapon. A schematic representation of the asymmetric electron flow is shown in Figure 2.8. Such an asymmetric current distribution will give rise to a predominantly vertically-polarized, radiated electromagnetic field.

In the case of a surface nuclear burst, the current distribution is far more asymmetric than that shown in Figure 2.8. In general, the greatest asymmetry of current flow occurs when the burst is at the surface of an infinitely-absorbing, infinitely-conducting ground. From the viewpoint of EMP generation, it is reasonable to assume that the earth infinitely absorbs gamma rays, since the mean free path of gamma radiation at the earth's surface is about 1/1000 of its mean free path in a standard atmosphere. The conductivity of the earth, also, is very high, although not infinite. Little error in the EMP field intensity determination results, however, if the earth's conductivity is assumed infinite. An illustration of the surface burst as the EMP source is shown in Figure 2.9.

For the high-altitude burst, Compton electrons are deflected by the earth's magnetic field in the deposition region as shown schematically in Figure 2.10. This deflection produces a large component of Compton current in non-radial directions. The non-radial component of the current produces a radiating electromagnetic field of large amplitude covering a large geographic area.

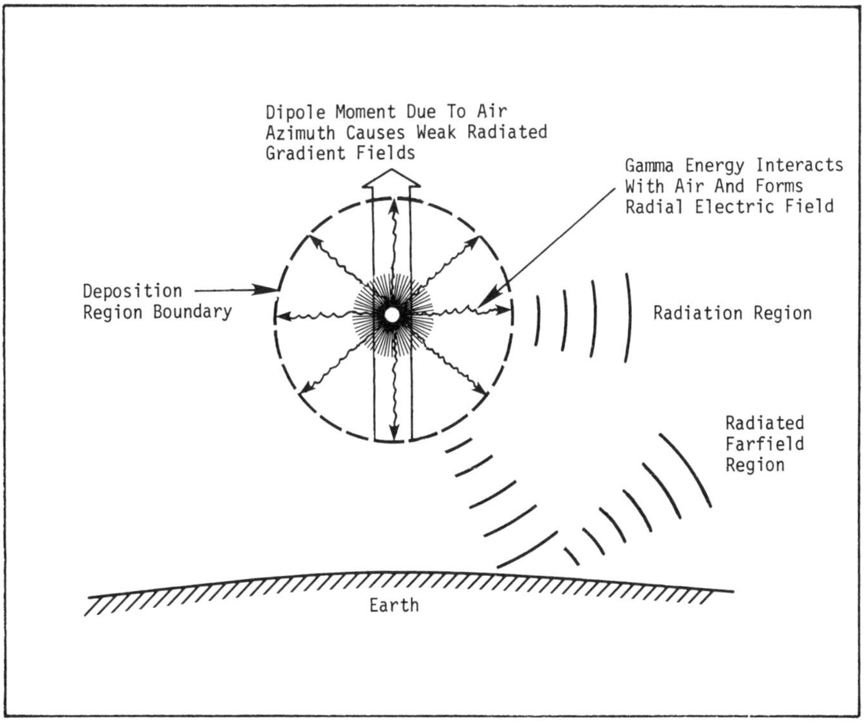

Figure 2.8 - The Earth's Surface Disturbs The Spherical Symmetry Resulting EM Radiation at Far Zone.

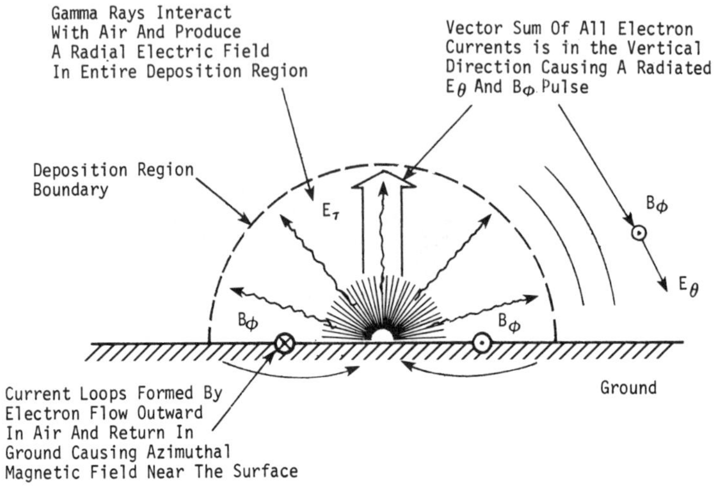

Figure 2.9 - Concept Of An Equivalent Vertical Dipole Formation From A Surface Nuclear Burst.

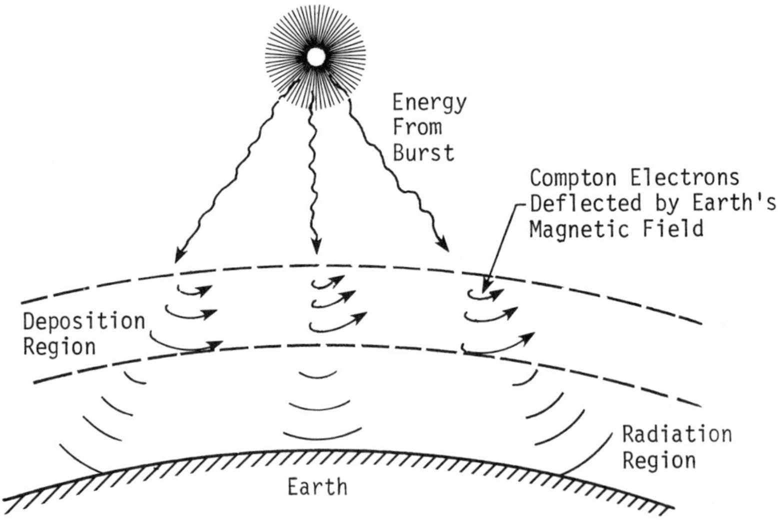

Figure 2.10 - Illustration of an Assymetry of Electron Flow From a High-altitude Burst in the Presence of the Earth's Magnetic Field.

2.1.3 Electromagnetic Field of EMP

When dealing with EMP generation and the determination of its field strength, it is helpful to have some basic understanding of how such a field may be created. Although calculation of the EMP field starting with the time-varying gamma radiations from different weapons is beyond the scope of this text, one may start with Maxwell's equations to develop a basic understanding of the EMP generation concept.

The EMP field is governed by Maxwell's two curl equations involving the electric field intensity vector $\vec{E}$ and the magnetic field intensity vector $\vec{B}$.

They are:

$$\nabla \times \vec{E} = -\frac{\partial \vec{B}}{\partial t} \qquad (2.1)$$

$$\nabla \times \vec{B} = \mu\vec{J} + \mu\varepsilon\frac{\partial \vec{E}}{\partial t} \qquad (2.2)$$

where:
$\vec{J}$ is the current density vector
μ is the permeability of the medium
t is real time
ε is the permittivity of the medium

The current density term $\vec{J}$ in Equation 2.2 can be broken into two components, the driving current $\vec{J}_d$ and the conduction current $\sigma\vec{E}$, σ being the conductivity of the medium. Further, the driving current, which is produced by the interaction of high-energy photons with the surrounding media, can be divided into three terms in accordance with the processes which create them.

Thus:

$$\vec{J} = \vec{J}_d + \sigma\vec{E} \qquad (2.3)$$

and

$$\vec{J}_d = \vec{J}_c + \vec{J}_{pe} + \vec{J}_{pp} \qquad (2.4)$$

where:
$\vec{J}_c$ is the Compton current
$\vec{J}_{pe}$ is the photoelectric current
and $\vec{J}_{pp}$ is the pair-production current

The Compton current, photoelectric current and pair-production current are, respectively, those resulting from the electrons due to Compton effect, photoelectric effect and pair-production effect, as discussed earlier in connection with the initial gamma radiations. The most important contribution toward $\vec{J}_d$, as noted earlier, is $\vec{J}_c$, unless one is concerned for a region where

X-rays are important, in which case the $\vec{J}_{pc}$ term must be considered. The term $\vec{J}_{pp}$ is usually negligible, because of the high energies required for significant amounts of pair-production and because the positive and negative currents cancel, except in the presence of a magnetic field. The pair-production mechanism cannot be ignored, however, since it affects the ionization rate and σ as it appears in Equation 2.3.

The basic process for the creation of the Compton current, $\vec{J}_c$, is already shown in Figure 2.2. The gamma radiation from a burst interacts with the air molecules, and, by means of the Compton process, creates high-energy electrons. Although the flow of free electrons generated or scattered by Compton effect is somewhat random, Compton electrons move, on the average, primarily in the radial direction, and, by their outward movements, create a Compton current along their path. In the process, these electrons lose energy to the surrounding air molecules through inelastic collisions. The energy from these collisions goes into the liberation of electrons from air molecules in a quantity termed the ionization rate. The creation of the secondary electrons e_s results in a conducting plasma around the burst point. The conductivity of the plasma is what is denoted as σ in Equation 2.3. The prompt gamma ray pulse increases rapidly to a peak value in about 10^{-8} second, and the Compton current varies with time in a similar manner.

To estimate the Compton current, one may consider again the radiation of gamma rays from a nuclear burst. As the gammas travel through the air, inelastic collision with air molecules occurs, causing a part of the gamma ray flux to be lost. This loss is in the form of $(e^{-R/\lambda})$, where R is the distance travelled by the gamma ray and λ is the mean-free path in air. In addition to this exponential attenuation, there is a loss of gamma flux in the form of $(1/4\pi R^2)$ due to geometric spreading. Thus, if the gamma flux at the source is represented as S(t), at a location R meters away, the direct-beam gamma flux becomes

$$f_\gamma(R,t) = \frac{e^{-R/\lambda}}{4\pi R^2} S(t) \tag{2.5}$$

λ being measured in meters.

Equation 2.5 denotes the direct-beam gamma flux at R. The total gamma flux at R also includes contributions from scattered gammas which are lost from the direct beam, but eventually reach R. The calculation for this scattered gamma flux is involved and is outside the scope of this text, since the development of a basic understanding of the EMP mechanism, instead of an exact calculation, is attempted here.

The flux of Compton electrons at R can be determined by finding how many Compton collisions are made within a certain volume and then considering how many of these collisions contribute to the electron flux at the location of interest. To determine the number of Compton collisions, the gamma flux can be divided by the Compton-gamma mean free path, λ_c for air. This gives Compton electrons within a given volume. However, since all such electrons do not create a flux at the point of interest, this expression must be multiplied by a factor R_e representing the mean forward range of the electron. The Compton electron flux at R resulting from the Compton effect, then, can be written as

$$f_c(R,t) = \frac{R_e}{\lambda_c} f_\lambda (R,t) \tag{2.6}$$

Finally, one obtains

$$|\vec{J}_c(R,t)| = q f_c(R,t) \tag{2.7}$$

q being the electric charge of an electron, since the electric current is just the charge in motion.

One needs to determine the conductivity of the medium in addition to the driving current density $\vec{J}_d$ to evaluate $\vec{E}$ and $\vec{B}$ components of the EMP field. This conductivity is due to the interactions of gamma and other nuclear radiations with the air and materials surrounding the location of burst and can be determined by summing the contributions from various charged species. Thus, one may write

$$\sigma = \sum_i q_i \mu_i n_i \tag{2.8}$$

where: q_i is the charge of the ith species
μ_i is the mobility of the ith species
and, n_i is the number density of the ith species.

For the purpose of determining σ, all charged particles can be grouped as secondary electrons, positive ions and negative ions. To determine the number density, all possible reactions involving the constituents of the air and output of the nuclear weapon need to be considered. Finally, the mobility is determined from its known functional relationship with the electric field, air density in the region of the burst, and water vapor content of the air. The computation of the conductivity as required for the solution of Maxwell's equations to evaluate the EMP field, particularly for various weapon yields and heights of bursts, will not be attempted here. Equations 2.1 through 2.8 merely sketch the analytical process which can be used to assess the EMP characteristics.

2.1.4 Charge Separation Model of EMP Generation

As mentioned earlier, the Compton current, which initially varies with time as the gamma ray phase, reaches its peak value in about 10^{-8} second. When the electrons are driven radially outward by the flux of the gamma rays, the atoms and molecules from which such electrons have been removed, i.e. the positive ions, travel more slowly because of their heavier masses. This results in a partial separation of charges with an inner shell of positive charges and an outer shell of negative charges. The separation of charges gives rise to a large, local, radial E field. Subsequently, the recoil electrons drift back toward the positive-charge shell and reduce the radial E field.

Using the charge separation model of EMP generation and assuming a spherical symmetry, one may write from Equation 2.2

$$\varepsilon \frac{\partial E_r}{\partial t} = J_c - \sigma E_r \tag{2.9}$$

where E_r is the magnitude of the radial electric field
t is real time
J_c is the magnitude of the radial current of the Compton recoil electrons
and σ is the air conductivity that results from the ionization produced by the Compton electrons.

Equation 2.9 follows from the situation that there is no magnetic field, and J_c is directed radially inward because the electrons move radially outward.

At early times, σE_r is much less than J_c, and

$$E_r \simeq \int \frac{J_c}{\varepsilon} \, dt \tag{2.10}$$

A little later in time, σ becomes large in a region close to the burst and the conduction current σE_r becomes much larger than $\varepsilon \frac{\sigma E_r}{\partial t}$. This leads to a saturation condition, where

$$E_r \simeq J_c/\sigma$$

The charge separation model of the EMP generation does not produce a radiated EMP because of the symmetry of electron flow as discussed earlier, although it can produce a large, local, radial electric field. The current distribution around the burst, however, is asymmetric most of the time. The air/ground interface for the surface burst and nonuniform gradient of the atmosphere with height make the net current density vector directed vertically, although there will usually be an azimuthal symmetry around this vertical direction. Far away from the burst, then, the current density can be characterized as a vertical line of current or an electric dipole.

The dipole moment P for such a case will be equal to QS, Q and S being, respectively, the charges and their effective separation distance. Also, the dipole moment will be related to the current I, such that I is proportional to dP/dt. Finally the electric field of EMP will be proportional to dI/dt or dP^2/dt^2. Figure 2.11 shows the time-varying nature of P, I and the corresponding electric field.

The basic principles of the EMP generation for the high-altitude, air and surface bursts are similar, although there are characteristic differences in the radiated EMP for such cases, as will be discussed further in the following subsections. The mechanisms of the generation of the internal EMP, system-generated EMP and underground EMP, however, are substantially different from those due to the surface and above-surface nuclear bursts. For example, the internal EMP is produced by electron emission from the walls of the system enclosure. The electrons emitted in such a case can be either Compton electrons, induced primarily by gamma rays, or photoelectrons, produced primarily by X-rays, or both. In the system-generated EMP, the nuclear radiation incident on a vehicle in space (such as a satellite), causes the

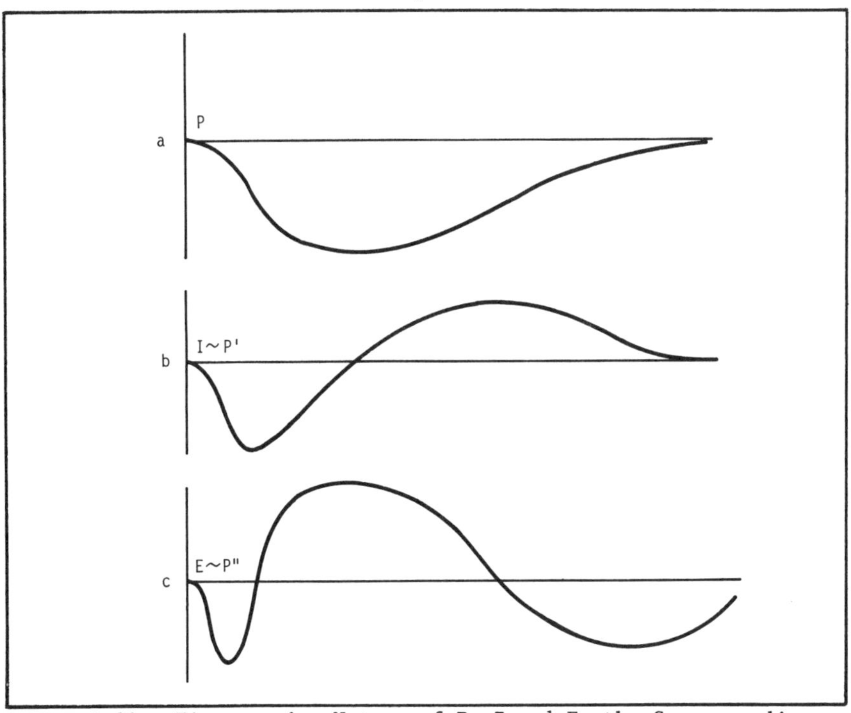

Figure 2.11 - Time-varying Nature of P, I and E, the Corresponding Electric Field.

emission of electrons from the outside surface of the vehicle. If the projected cross-section is not symmetrical, then more charge can be lost from one end than another, resulting in a large surge-current along the vehicle structure required to redistribute and equalize the charge distribution over the outside surface. If the system cables are not well shielded, these surge currents and external fields will couple energy into the electrical wiring of the system, as the system-generated EMP.

When a nuclear explosion occurs underground, the expansion of the nuclear radiation is contained. An EMP can still be generated in such a case by the magnetohydrodynamic effect where the earth's magnetic field is displaced by the products of the underground nuclear detonation. Still another mechanism of EMP generation from underground explosion involves large gamma-induced currents on the metallic casing which usually lines the vertical shaft of the test hole. These large currents may radiate like a cylindrical antenna, except high frequencies in the radiated field are more attenuated than lower ones since the casing is immersed in a conducting earth.

A discussion of the characteristics of EMP generated from the underground burst and from bursts at various altitudes above the earth's surface will now be in order.

2.2 EMP for High-Altitude Bursts

A high-altitude burst is defined as a detonation at an altitude of above 30 kilometers. For such a detonation, gamma rays moving in the upward direction enter an atmosphere where the air density is so low that the rays travel great distances before being absorbed. The gamma rays emitted from the explosion in the downward direction, however, encounter a region where the atmospheric density is increasing. Then, downward gamma rays interact with the air molecules and atoms to form a deposition region which, as mentioned earlier, is the source of EMP. For the high-altitude burst, the thickness and extent of the deposition region, which is a function of the yield and altitude of burst, is shown in Figures 2.5 and 2.6. The deposition region is thicker immediately below the burst than at larger distances, because the gamma-ray intensity decreases with distance from the burst point. Since the gamma rays pass through air of increasing density as they travel toward the ground, most are absorbed in a layer between the altitudes of about 15 to 65 kilometers.

The flow of electrons from the burst as generated by the Compton effect becomes asymmetric due to the varied atmospheric density with height. Such an asymmetry alone can cause a radiated EMP. In the case of the high-altitude burst, the EMP radiation at the far field can also be caused by a different mechanism. For example, Compton electrons from the high-altitude burst are forced to follow curved paths of the geomagnetic field lines. In doing so, they are subjected to radial acceleration; and the ensemble of turning or spiraling electrons, whose density varies with time, emits coherent electromagnetic radiation. The high-frequency content of the EMP thus produced is significantly more than that present in EMP arising from local asymmetry.

EMP from a high-altitude burst does not radiate solely in the direction downward from the source region. It also radiates from the edges and at angles other than the vertical beneath this region. The effect of the earth's surface on the high-frequency EMP is to extend it to the horizon and not beyond, since the propagation of electromagnetic waves at high frequencies is more or less limited to the line-of-sight path. The lower frequencies, however, will extend beyond the horizon, by the well-known diffraction process. The radius of the high-altitude burst effect is determined by the tangent to the earth's surface down from the burst point. Thus, the radius of the effect can be 1,800 to 2,400 Km, depending on the altitude of burst. Within this entire region, the peak-field intensity is essentially the same (within a factor of two) from the earth's surface up to about 20 Km altitude, except for a small quiet zone around the burst-point magnetic field line. A more illustrative description of the EMP ground coverage and the variation of EMP peak field for various altitudes of bursts, assuming a high-altitude detonation above a region in the State of Kansas of the United States of America, is shown in Figure 2.12. The far-reaching EMP impact probability is evident from this figure.

The peak-field intensity from a high-altitude nuclear burst can be very large. For a high-yield nuclear detonation, EMP could have a peak-field intensity of 100,000 volts per meter. It may be remarked that the average radio-frequency-field intensity considered safe for a human being is about 200 volts per meter, although one can withstand a higher peak value of the field with a lower average value.

The pulse shape, or waveform, of EMP resulting from a high-altitude burst can be approximately represented by the difference of two time-varying exponentials, as shown in Figure 2.13a. The rise time can be on the order of 10 nanoseconds, due to the comparable "unfolding time" of the fission of uranium (U^{235}). The decay time depends on the duration of the nuclear weapon gamma rays, a few hundred nanoseconds being the time by which the field intensity decays to a factor of 1/e.

The frequency spectrum of EMP originating from the high-altitude burst is extremely broad, extending from the extremely low frequencies (ELF) up to the ultra-high frequencies (UHF) and beyond. The shape of the expected spectral distribution of the E field of the EMP is shown in Figure 2.13b.

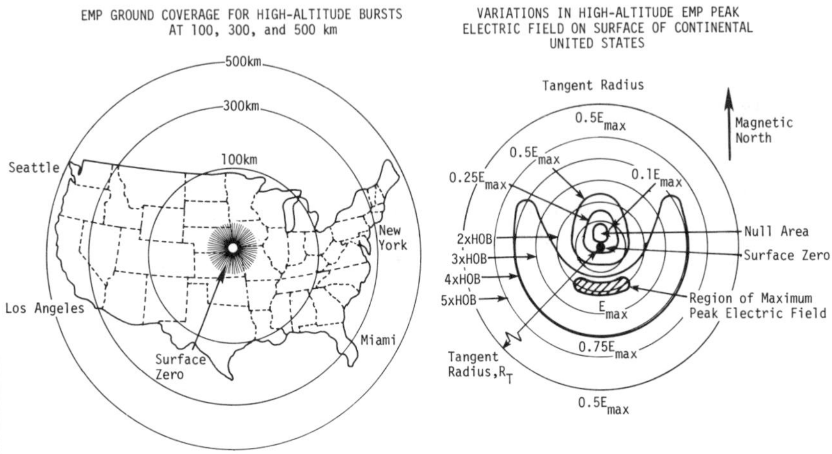

Adapted with the permission of the Defense Nuclear Agency

Figure 2.12 - Ground Coverage and Peak Field Variation of EMP for Various High-altitude Detonations.

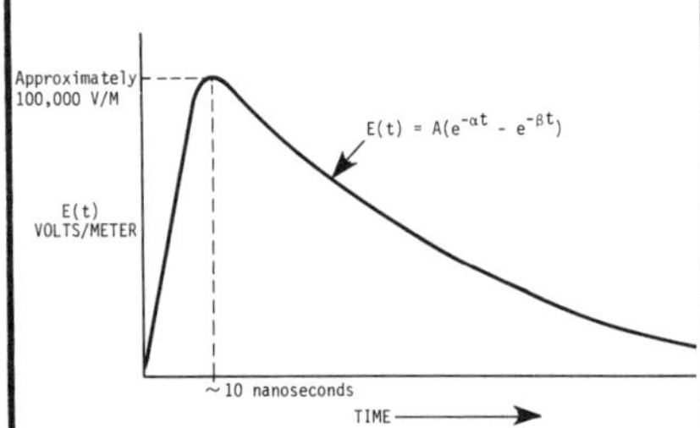

Figure 2.13a - The Pulse Shape or Waveform of EMP Resulting from a High-altitude Burst.

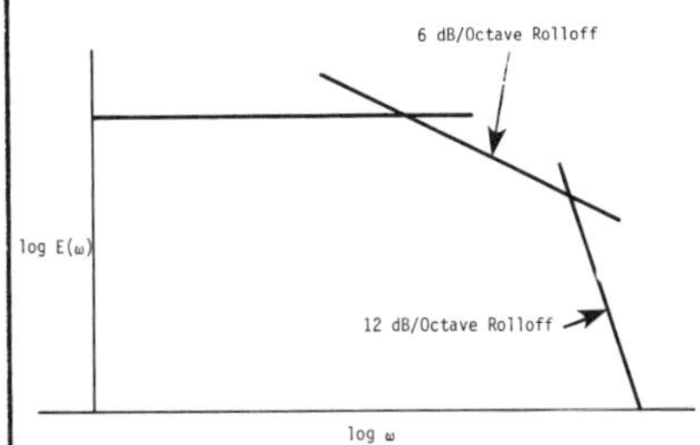

Figure 2.13b - Expected Spectral Distribution of the E Field from a High-altitude Burst.

2.2.1 EMP Characteristics for Nuclear Detonation in Space

As early as 1961, Karzas and Latter attempted to obtain an order of magnitude estimate of the EMP characteristics for the space nuclear detonation. They attributed the EMP radiation to the asymmetric distribution of Compton electrons caused by either the design of the nuclear device or the existence of an X-ray shielding material surrounding the device. The exponential buildup of gamma rays is assumed as the form $\alpha e^{\alpha t}$, where α is assumed to be 10^8 $\sec^{-1}$. This exponential buildup is assumed to continue until a substantial fraction of the full yield has been produced, at which time the rate reaches a maximum. Thereafter, the gamma-ray generation proceeds at a lower rate, decreasing according to the equation

$$\sigma\beta e^{-\beta(t-T)} + \alpha T \tag{2.11}$$

where
$\sigma \leq 1$
T = time at which the peak gamma-ray output occurred
$\alpha = 10^8$ $\sec^{-1}$
$\beta << \alpha$

It was assumed that 0.03% of the weapon yield appeared as gamma rays with an average energy of 1 Mev and that 7.5×10^{21} gamma rays were produced per kiloton of yield.

The EMP effect of the expected Compton electrons was found to be independent of the nuclear yield, and the peak-electric-field strength in the far-field was determined to be

$$E_{pk} \simeq 10^4 \frac{1}{R^{3/2}} \text{ volts/meter} \tag{2.12}$$

where R is the range from the burst in kilometers. The pulse duration and bandwidth of the EMP were estimated to be $R \times 10^{-10}$ seconds and between 10 and 100 MHz, respectively.

As noted above, the 1961 analysis of Karzas and Latter assumed the EMP generation from a space burst due to the asymmetric distribution of Compton electrons. Later, it became apparent that an interaction with the earth's magnetic field was possible. This lead to an atmospheric nuclear test and numerous studies to understand the test data. Many of these studies supported the assumptions of Karzas and Latter. In one such study, undertaken in 1964, Karzas and Latter extended their earlier theory to obtain estimates of the maximum detection range of a nuclear detonation in space and concluded that such events could be detected to a distance R_{max}, where $R_{max} \simeq 10\sqrt{Y}$ kilometers, Y being the nuclear weapon yield in kilotons. On this basis, a high-altitude detonation of a megaton-yield nuclear weapon can be detected further than 300 million kilometers from the source - an impressive distance!

2.3 EMP from Ground and Near-Ground Bursts

As in the case of high-altitude bursts, the primary source of the EMP from ground bursts is gamma radiation. There are some significant differences, however, in the generation mechanism, particularly those arising from the asymmetry of electron flow. More specifically, for a ground burst, gamma rays travelling in a generally downward direction are readily absorbed in the upper layers of the earth's surface, and there is essentially no charge separation or electric field in this direction. The gamma rays moving upward and outward, on the other hand, produce ionization and charge separation in air, as they leave behind the positively-ionized air molecules near the burst point. The initial effect of a ground burst as an EMP source is shown schematically in Figure 2.14.

The gamma rays from the ground burst interact with the air within a few hundred meters at sea level. Each Compton electron photon may ultimately cause the generation of 30,000 electrons. These electrons, being more mobile than the more massive positive ions can move out to a distance of 1 to 10 kilometers. The large charge separation leads to a large radial electric field between the burst point and the edge of the ionosphere. The net effect of the generally upward movement of the charges is to generate a horizontal magnetic field. By virtue of Ampere's law, the line integral of the magnetic field around the electron flow is equal to the effective vertical current. In addition, as the electrons tend to move toward the ground through the easiest path to get back to the positive charges left behind, another component of the magnetic field is generated at the transverse plane of the current flow, as shown in Figure 2.15. This component of the magnetic field is also a source of EMP.

The electric field produced in a surface burst is very strong, but the radiated field falls off very rapidly with distance (as $1/R^3$), at first from the deposition region, and then somewhat less rapidly afterwards (as $1/R^2$ and $1/R$).

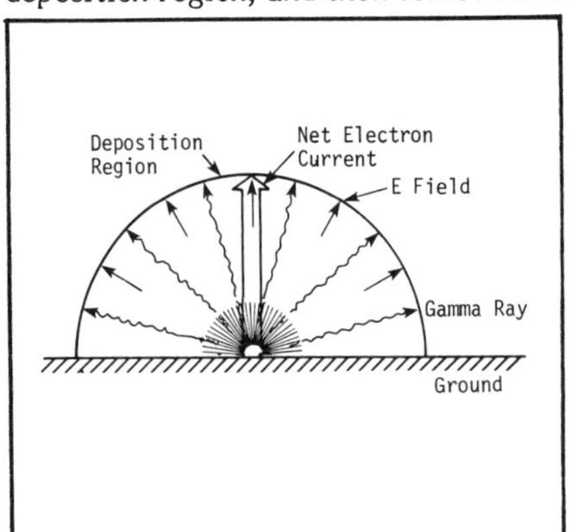

Figure 2.14 - Schematic Representation of the EMP in a Surface Burst.

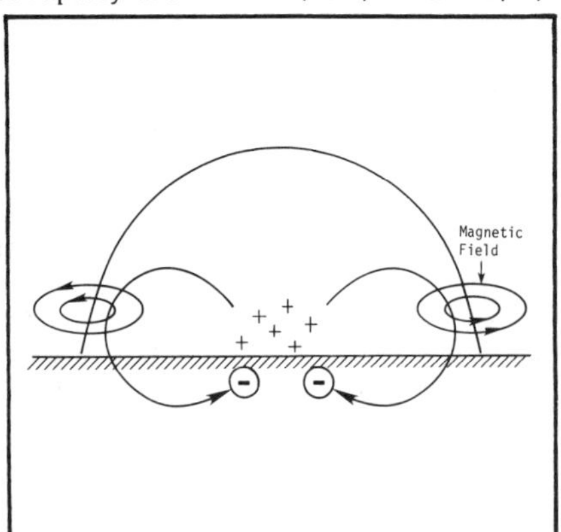

Figure 2.15 - Generation of Azimuthal Magnetic Fields by Charges Returning to Ground.

The potential hazard to electric and electronic equipment from EMP will, thus, be greatest within and near the deposition region, extending to 3 to 8 kilometers, depending on the weapon yield. In this area, the structures in which an equipment is housed may suffer damages, especially from high-yield explosions, unless they are blast-resistant.

At distances away from the deposition region, the variation of peak-radiated EMP field from a surface burst varies as

$$E = (R_o/R)\,E_o \tag{2.13}$$

where E_o is the peak-radiated field at the radius R_o of the deposition region, and E is the peak field at the surface distance R. The peak-radiated field at the boundary of the deposition region is ten to a hundred times stronger in a direction along the earth than for a similar air burst. E_o is of the order of several kilovolts per meter.

Underneath the ionized region resulting from the ground burst, the greatest concern from EMP viewpoint is the magnetic field along the direction of the earth. In this region, the vertical electric and horizontal magnetic field components are not related through the usual free-space impedance of 377 ohms; and, in fact, the E/H ratio varies with time. Representative variation of H field with time, generated from a ground burst, is shown in Figure 2.16.

If the detonation is not at the zero altitude, but at a few hundred feet, gamma rays can impinge on the ground over a larger area. For such a case, the conductivity of the ground is greatly enhanced by the ionization, and very large currents can flow into the ground. This generates magnetic fields near the earth's surface which are somewhat larger than the zero-altitude case and which are oppositely directed.

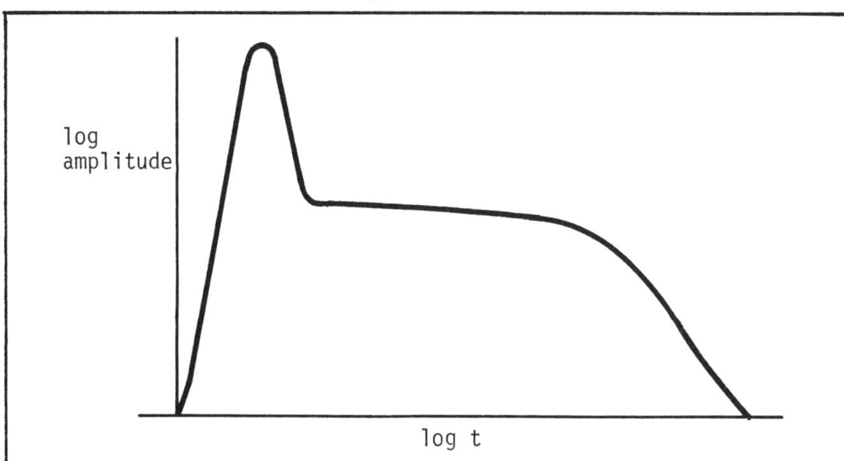

Figure 2.16 - Variation of Magnetic Field With Time, Following a Ground Burst

Although EMP can be generated from underground nuclear detonations and other mechanisms as discussed in following subsections, EMP threats to most ground and airborne systems are primarily due to high-altitude, near-surface, and surface nuclear bursts. A qualitative comparison of the EMP environment, particularly with respect to EMP rise time, peak altitude, and energy spectrum, resulting from the high-altitude, near-surface (air), and ground bursts, is shown in Table 2.1.

Table 2.1—EMP Environment Variations—A Qualitative Comparison

Burst Product \ **Location of Burst**	**Surface (0-2km)**	**Air (2-20km)**	**High Altitude (40km)**
Pulse Risetime	50ns (E_r) 3μs (E_{pk})	50ns (E_r) 3μs (E_{pk})	5ns (E_{pk})
Peak Amplitude	100kV/m (E_r) ~1kV/m (E_{pk})	10kV/m (E_r) ~.1kV/m (E_{pk})	50-100kV/m
Energy Spectrum	<1MHz	<1MHz	.01-100MHz
Coverage: Deposition	R<6km	5-15km	————
Radiated	R<200km	15-100km	1000km

2.4 EMP from Underground Burst

When a nuclear explosion occurs below ground, the expansion of the ionized region is contained. A number of mechanisms may still exist for the generation of EMP under such a circumstance.

One mechanism, for example, involves large gamma-induced currents on the metallic casing which typically lines the vertical shaft of the test hole. These large currents radiate in a manner similar to a vertical electric current element, or dipole, immersed in the conducting earth. Since the field from such a vertical dipole is proportional to $\exp(-i\beta_1 d)$ where d is the effective depth of the dipole, and $\beta_1^2 = \omega^2\mu\varepsilon_1 - j\omega\mu\sigma$, ε_1 and σ being, respectively, the dielectric constant and conductivity of the earth, the high-frequency components of the EMP generated at the underground source do not radiate well at any reasonable distance from the source.

Another mechansim of EMP generation from an underground source is the direct radiation from a plasma resulting from a nuclear detonation but contained inside a cavity, as shown in Figure 2.17. For this type of source, it may be assumed that the explosion creates an infinitely conducting, expanding spherical plasma which interacts with the earth's magnetic field. This magnetic field is frozen in the plasma as it expands. Thus, if the expansion is assumed to begin from a point, no magnetic flux will exist inside the volume containing the plasma.

The physical mechanism for canceling the earth's magnetic field, say $\vec{B}_o$, inside the plasma is a surface current which is induced by $\bar{B}_o$. Hence, the plasma, in this case, may be replaced, for the purpose of determining the EMP field, by an equivalent current density $\bar{J}$ in the earth. This is more convenient than solving directly the difficult moving boundary value problem.

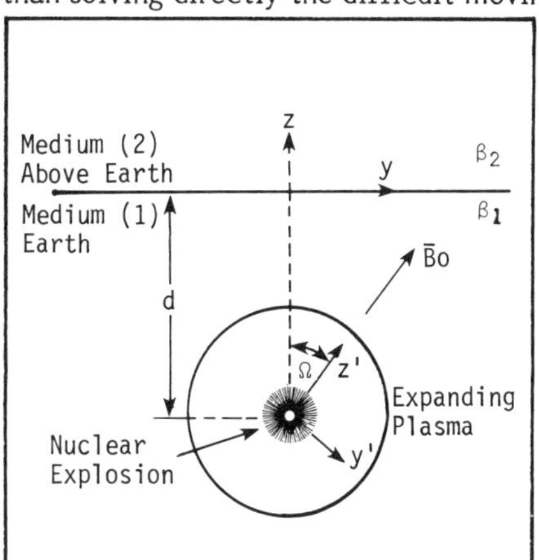

Figure 2.17 - EMP Generation Due to Expanding Plasma from an Underground Detonation.

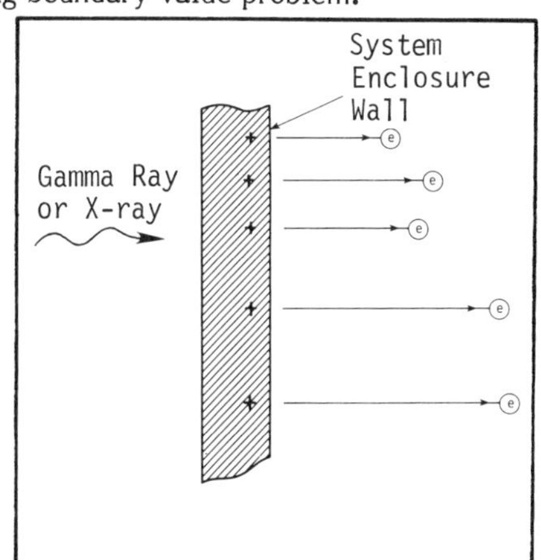

Figure 2.18 - Electron Emission and Resulting Fields from Radiation Impinging on Enclosure Wall.

Following Sommerfield's method of solving half-space boundary value problems, Stogryn and Ghose developed an expression of the electric field resulting from an explosion inside an underground cavity of any arbitrary depth d below the earth's surface. Like the field resulting from any underground source, this field is seen to be proportional to $\exp(-j\beta_1 d)$. In addition, the field is directly proportional to B_o and $\sin\Omega$, Ω being the angle between the earth's magnetic field and the local vertical. Because of the exponential factor $(-j\beta_1 d)$, the contribution of high-frequency components of the EMP field becomes progressively small with d.

Still another mechanism of EMP generation from an underground source involves the radiation from an ionized column formed when the plasma resulting from a nuclear explosion is allowed to escape the underground hole. In effect, this column carries currents which may radiate as a vertical radiating antenna immersed in the earth.

Because of the inherently inefficient coupling of the source energy to the electromagnetic radiation field, the effect of EMP from an underground nuclear burst is not often of concern from the viewpoint of equipment or system vulnerability.

2.5 Internal EMP

When the enclosure wall of a system such as a satellite or a missile is subjected to a very large ionizing environment consisting of gamma rays or X-rays, electrons are emitted from the wall as illustrated in Figure 2.18. These electrons can be Compton electrons induced primarily by gamma rays, or photoelectrons produced by X-rays, or both. Although electrons are generated throughout the thickness of the wall as the radiation penetrates the wall material, only the electrons close to the inner surface of the wall have sufficient energy to escape the surface and pass into the space behind the wall. This is because the range of the electrons (that is, the maximum distance they can travel before loosing their energies and therefore coming to rest) is very short in solid materials.

The electrons emitted from or near the surface wall do not come out of the surface in a direction perpendicular to the surface, even when the incident radiation is normal to the wall. The emitted electron direction can be expressed in terms of a distribution of electrons over all possible angles. The net effect of the gamma-ray or X-ray radiations, then, is that the electrons of various energies proceed inward, away from the wall, in various directions.

For each electron emitted, there is a positive ion left behind in the wall material and an electric field is produced as a result of the charge separation. This field is referred to as the "Internal EMP," or "IEMP," and it may affect the system operation, depending on the sensitivity of the system.

The internal EMP field exhibits a Coulomb-force* that tends to decelerate the electrons. If the space behind the wall is a vacuum or near vacuum, as in the case of a satellite in space, for example, the electrons having sufficient energy to overcome the Coulomb-force may continue across the void until they strike some solid material. If the material ultimately stopping the electrons is an electrical conductor, the electrons can flow from the stopping point back through the metallic structure to the positive charges left behind in the enclosure wall. If, however, the solid material is a dielectric instead, current flow becomes limited.

During the time interval when the electrons are traversing the void, or cavity, between the enclosure wall and the stopping wall, very large, short-duration electromagnetic fields are generated in the cavity. These fields may even be magnified at some point or points inside the cavity due to cavity resonances, depending primarily on the cavity geometry, dimensions, the direction of excitation, and cavity-wall conductivity. The cavity mode of IEMP generation is illustrated in Figure 2.19. Here gamma rays produce Compton electrons inside the cavity and excite the cavity, setting up electromagnetic fields inside the cavity. Typical directions of the electric field component E and the magnetic flux density B of the field excited inside the cavity are shown in the figure.

To predict analytically the magnitude of the IEMP effect in a particular case, it is necessary to have a good model for predicting the electron emission from the nuclear radiation. This depends on the geometry, the direction of the incident radiation, energy spectrum of the radiation, and the type of materials involved. In addition, it is necessary to have a model for predicting the electromagnetic fields produced by these electrons, which requires simultaneous solution of Maxwell's equations with plasma equations. A detailed analysis of the IEMP generation mechanism for various geometries of system wall structures and incident radiations will not be attempted here. We may note here, however, the results of some analyses on the subject to indicate the nature of the system vulnerability problems which may be encountered due to the IEMP.

---

* Here, the Coulomb force is due to the force of attraction between opposite charges.

2.5.1 Characteristics of IEMP

Baum analyzed the IEMP fields for many simple geometries, including cylindrical and spherical configurations, in terms of the incident gamma radiation. When the enclosure wall is a cylinder, for example, with its radius R much smaller than its length L, as shown in Figure 2.20, and the gamma radiation perpendicular to the cylinder axis, the radial field is given by

$$E_r \simeq -0.5 \times 10^{-5}\ \gamma r \quad \text{volts/meter} \tag{2.14}$$

the field at the center of the cylinder being zero. The maximum E-field occurs at r = R and is given by:

$$|E|_{max} \simeq 0.5 \times 10^{-5}\ \gamma \cdot R \quad \text{volts/meter} \tag{2.15}$$

The voltage distribution, given by:

$$\int_r^{R_o} E_r\, dr$$

is,

$$V(r) = -0.25 \times 10^{-5}\ (R^2 - r^2)\,\gamma \text{ volts} \tag{2.16}$$

and the maximum potential difference which exists between r = 0 and r = R is

$$|V|_{max} = 0.25 \times 10^{-5} \cdot \gamma R_o^{\,2} \quad \text{volts} \tag{2.17}$$

The magnetic field produced for such a case is given by:

$$H_z = -2 \times 10^{-8} \cdot \gamma r \sin\phi \quad \text{amps/meter} \tag{2.18}$$

$$|H_z|_{max} = 2 \times 10^{-8}\ \gamma R \quad \text{amps/meter} \tag{2.19}$$

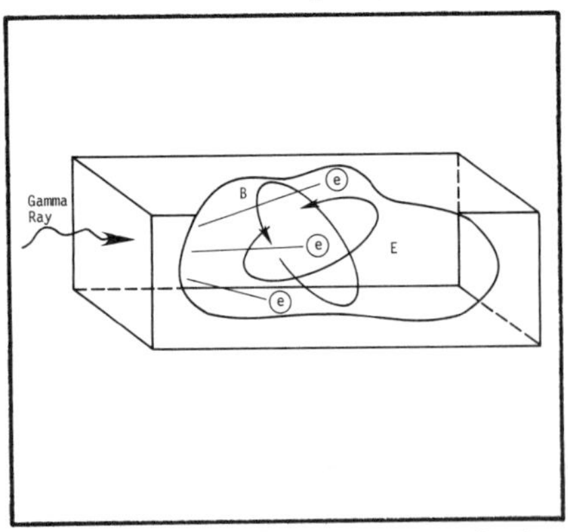

Figure 2.19 - Illustration of the Cavity-mode of IEMP Generation.

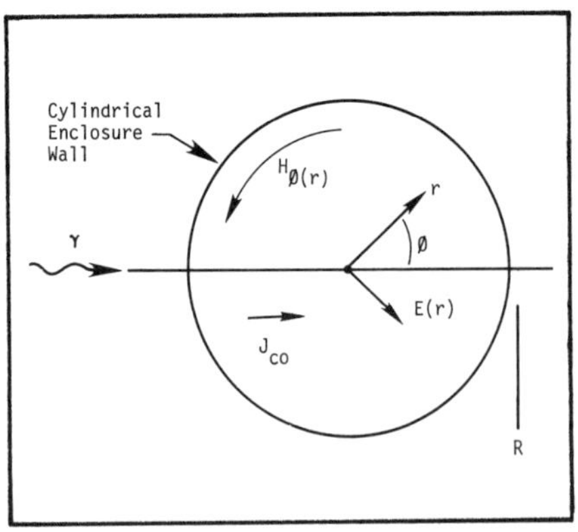

Figure 2.20 - Internal EMP Field Components.

ϕ being the azimuthal angle as shown in Figure 2.19. The energy is the electric field, given by

$$\tfrac{1}{2}\varepsilon_o \iiint_V E_r^2 \, dv$$

ε_o being the free-space dielectric constant, is

$$U_e \simeq 1.8 \times 10^{-22} \gamma^2 R^4 L \text{ joules} \tag{2.20}$$

Similarly, the energy in the magnetic field is

$$U_m \simeq 2 \times 10^{-22} \gamma^2 R^4 L \quad \text{joules} \tag{2.21}$$

When the Compton current is parallel to the cylinder axis, as shown in Figure 2.19, the E-field remains the same and the magnetic field has the same magnitude as before but has a different direction, as shown in Figure 2.21. The energy in the magnetic field also becomes different and is given by

$$\bar{U}_m = 10^{-22} \gamma^2 R^2 L \quad \text{joules} \tag{2.22}$$

To obtain an estimate of the IEMP characteristics to be expected realistically, one may assume a value of $\gamma = 10^8$ (Routgens/sec). If for such a situation R = 1 meter and L = 1 meter,

$$|E_r|_{max} \simeq 0.5 \times 10^{-5} \gamma \cdot R = 500 \quad \text{volts/meter} \tag{2.23}$$

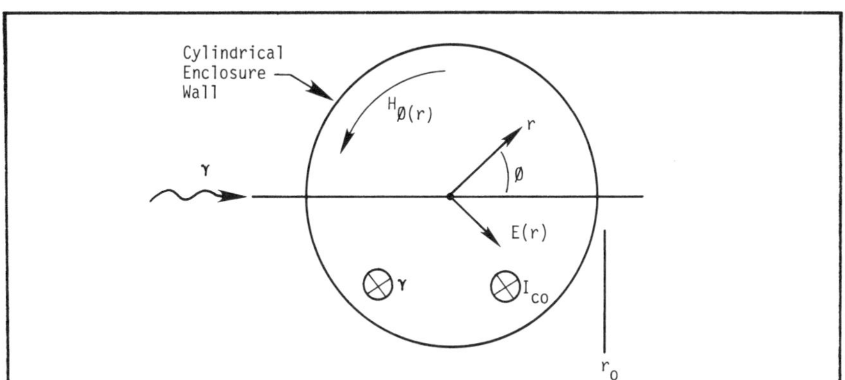

Figure 2.21 - Internal EMP Field Components Generated when Compton Current is Parallel to Cylinder Axis.

Similarly,

$$|V|_{max} = 0.25 \times 10^{-5} \gamma R_o^{2} = 250 \text{ volts} \quad (2.24)$$

$$|H_z|_{max} = 2 \times 10^{-8} \gamma R_o = 2 \text{ amps/meter} \quad (2.25)$$

$$|U_e| \simeq 1 \cdot 8 \times 10^{-22} \gamma^2 R^4 L \simeq 1 \cdot 8 \times 10^{-6} \text{ joules} \quad (2.26)$$

$$|U_m| = 2 \times 10^{-22} \gamma^2 R^4 L - 2.0 \times 10^{-6} \text{ joules} \quad (2.27)$$

In the case of a sphere of radius R, the radial electric field is given by:

$$E_r(r) \simeq -0.34 \times 10^{-5} \gamma r \text{ volts/meter} \quad (2.28)$$

with a maximum occuring at r = R.

The voltage distribution for such a case is given by:

$$V(r) \simeq -0.17 \times 10^{-5} \gamma (R^2 - r^2) \text{ volts} \quad (2.29)$$

and the maximum difference in potential which exists between the center of the sphere and R is:

$$|V|_{max} \simeq 0.17 \times 10^{-5} \gamma R^2 \text{ volts} \quad (2.30)$$

The magnetic field in this case is circumferential around the axis of the γ radiation and is given by:

$$H_\phi = -10^{-8} \gamma r \sin\phi \text{ amps/meter} \quad (2.31)$$

where ϕ is the angle between the vector $\vec{r}$ and the direction of the incident gamma ray. The maximum magnetic field at r = R and $\phi = \pi/2$ is:

$$|H_\phi|_{max} \simeq 10^{-8} \gamma R \text{ amps/meter} \quad (2.32)$$

The energies in the electric and magnetic field, for this case, are:

$$U_e \simeq 1 \cdot 3 \times 10^{-22} \gamma^2 R^2 \text{ joules} \quad (2.33)$$

$$U_m \simeq 10^{-22} \gamma^2 R^5 \text{ joules} \quad (2.34)$$

Again, assuming $\gamma = 10^8$ (Routgen/sec) and R = 3 meters:

$$|V|_{max} \simeq 1530 \text{ volts} \quad (2.35)$$

$$|H_\phi|_{max} \simeq 3 \text{ amps/meter} \quad (2.36)$$

$$U_e \simeq 11.7 \times 10^{-6} \text{ joules} \quad (2.37)$$

$$U_m \simeq 243 \times 10^{-6} \text{ joules} \quad (2.38)$$

Thus, it is seen that, although the energy content of IEMP is insignificant, there could be system susceptibilities due to high voltages and field intensities, depending on the system characteristics.

2.6 System-Generated EMP

In addition to EMP resulting from the interaction of gamma rays from a nuclear explosion with the atmosphere or ground, another mechanism of EMP generation is possible by the interaction of nuclear radiations or ionizations with various solid materials present in electronic systems. Such an EMP is referred to as the "System-Generated EMP," or "SGEMP." The generation mechanisms of SGEMP and IEMP are very similar, and sometimes they refer to the same physical phenomenon.

The theory of system-generated EMP mechanism is complex. In simple terms, it can be explained as follows. The solid material in an electronic system, or even in the shielding designed to protect the system from the external EMP, contains atoms which are heavier than those present in air. We have already seen that the interaction with gamma rays and high-energy X-rays will produce electrons by both Compton and photoelectric effects. These

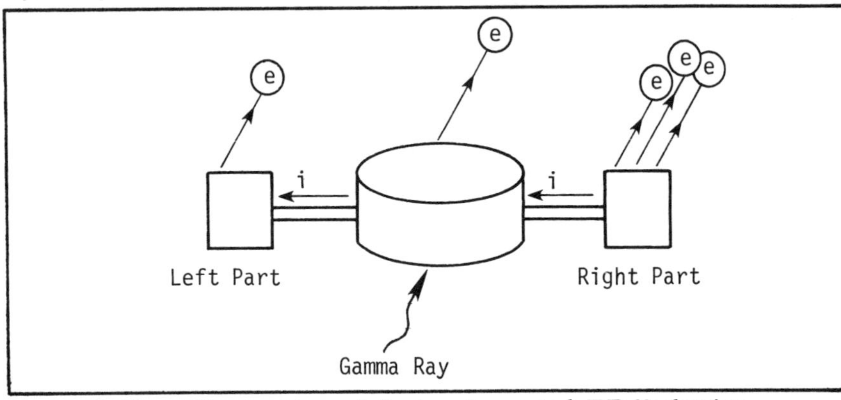

Figure 2.22 - Illustration of System Generated EMP Mechanism.

electrons can, in turn, interact with the solid materials to release more electrons by ionization. The latter electrons are called "secondary electrons." These secondary electrons, which are produced directly or indirectly, close to and on both faces of a solid material, have a velocity component perpendicular to the surface. As a result, an electric field is generated at the surface. SGEMP can also be generated if the projected cross-section of the system is not symmetrical, such that more charge can be lost due to electron emission from one end than another, resulting in a large surge current along the vehicle or system enclosure structure required to redistribute and equalize the charge distribution over the outside surface.

Such a mechanism is illustrated in Figure 2.22. Here it is assumed that, because of the asymmetry, more electrons leave the part of the system at the right than the part at the left. A current can then flow from right to left. If the system cables are not properly shielded, the surge current and the external field will couple energy into the system and may cause system vulnerability. Some of the electrons, as shown in Figure 2.22, will have sufficient energy to escape the vehicle altogether, leaving a static charge which will not decay for a long time when the vehicle is in space. The lower-energy electrons will be decelerated by the electric forces between the electrons and the positive ions left behind. The deceleration can bring the particle to a stop and then reaccelerate it back toward the space vehicle where it can bombard the vehicle.

The system-generated EMP is most important for electronic components in satellite and ballistic systems, above the deposition region, which would be exposed directly to the nuclear radiations from a high-altitude burst. The system-generated EMP can also be significant for surface and moderate-altitude bursts if the system is within the deposition region but is not subjected to damages by other weapon effects. This could occur for surface systems exposed to a burst of a relatively low-yield explosion, or for airborne systems and bursts of higher yield.

The electric fields generated near the walls of enclosures of systems by direct interactions of ionizing radiations can induce electric current in components, cables, ground wires, etc. Large currents and voltages capable of causing damage or disruption can be developed, just as with external EMP. Because of the complexity of the interactions that lead to the system-generated EMP, the effects are difficult to predict. The phenomenon of SGEMP appears to constitute an important consideration in satellite vulnerability.

2.7 References

1. Karzas, W. J., and Latter, R., *Electromagnetic Radiation from a Nuclear Explosion in Space*, Rand Report No. RM-2849-AFT, October, 1961.

2. Cowling, T. G., *Magnetohydrodyn Symp. 4*, New York Interscience Publishers, New York, N.Y.

3. Stogryn, A. P., and Ghose, R. N., *Generation of an Electromagnetic Pulse by an Expanding Plasma*, Radio Science Journal of Research, p. 147, Vol. 68D, No. 2, February 1964.

4. Baum, C. E., *Unsaturated Compton Current and Space Charge Field in Evacuated Cavities*, Air Force Weapons Lab, January 1965.

5. *DNA EMP Awareness Course Notes*, Edited by J. E. Bridges, ITT Research Institute, August 1971.

6. *Electromagnetic Pulse Sensor and Simulation Notes*, Vol. 1 - 10, Air Force Weapons Laboratory, April 1967 through 1972, AFWL EMP 1-1 through 1-10.

7. *EMP Protective Systems*, Department of Defense, Office of Civil Defense, November 1971, TR 61-B.

8. Gilinsky, V., and Pebles, G., *The Development of Radio Signals from a Nuclear Explosion in the Atmosphere*, Journal of Geophysical Research, Vol. 73, p. 405, 1968.

9. Karzas, W. J., and Latter, R., *Detection of Electromagnetic Radiation from Nuclear Explosion in Space*, Physics Review, Vol. 137, p. 1369, 1965.

10. Ricketts, L. W., et al, *EMP Radiation and Protective Techniques*, Wiley Sciences, 1976.

11. Erkkila, J. H., *Calculations of EMP from High Altitude Nuclear Detonations*, AFWL Report, April 1967.

12. Suydam, B. R., *Electromagnetic Signal from a Bomb Burst in Vacuo*, Los Alamos Scientific Laboratory, November 1961.

13. Karzas, W. J., and Latter, R., *The Electromagnetic Signal Due to the Exclusion of the Earth's Magnetic Field by Nuclear Explosion*, Rand Corp. Report No. RN-2890-PR, December 1961.

14. Bjorklund, R. F., *The Close-In Electric Field Generated by a Spherically Symmetric Nuclear Burst*, Rand Report RM-2955, April 1962.

15. Karzas, W. J., and Latter, R., *Electromagnetic Signals Produced by Low-Altitude Nuclear Explosions*, Rand Report RM 2884-PR, December 1963.

16. Gilinsky, V., *The Kompaneets Model for Radio Emission from a Nuclear Explosion*, Rand Report RM-4134, August 1964.

17. Baum, C. E., *A Technique for the Approximate Solution of EMP Fields from a Surface Burst in the Vicinity of an Air-Ground or an Air-Water Interface*, Air Force Weapons Lab, September 1966.

18. Longmire, C. L., *External System-Generated EMP on Some Types of Satellite Structure*, Mission Research Corp., August 1971.

19. Vittitoe, C. N., *Models for Electromagnetic Pulse Production from Underground Nuclear Explosions, Part I: Metallic Casing Effects*, Sandia Laboratories, April 1970.

20. Longmire, C. L., *On the Electromagnetic Pulse Produced by Nuclear Explosion*, IEEE Trans. AP-26, 1978.

21. Longmire, C. L., *EMP Environments for Surface and High Altitude Nuclear Explosions*, Mission Res. Inst., Third Symp. and Tech. Exhibition on Electromagnetic Compatibility, Rotterdam, May 1979.

CHAPTER 3

Modes of Entry of EMP Energy into Systems

Modes of entry relating to EMP refer to the modes or means of coupling EMP energy into a system. Usual modes of entry are through antennas, direct penetration of shields and enclosures and through slots and slits, doors and windows of the enclosures, etc. A coupling of energy may also take place through cables and transmission lines and utility lines. In addition, there could be hybrid modes involving coupling of EMP energy by more than one mode. For the vulnerability prediction due to EMP, at least the dominant coupling modes need to be investigated. These dominant modes can be determined by a physical inspection of the system of concern and by examining the sensitivity of components or subsystems from the viewpoint of performance and mission. Characteristics of various EMP coupling modes have been investigated in recent years, both experimentally and through theoretical analyses. Results of some of these investigations and the relative importance of various coupling modes are discussed in this chapter.

3.1 Importance of Coupling Mode Determination

In order for an electronic or weapon system exposed to the EMP field to be vulnerable, there must be a collection of EMP energy over a sufficient area, since such an energy received per unit area is usually small, except when the system is located within or near the deposition region. As noted earlier, however, the damaging voltages and currents in electrical circuits and systems can occur even when the collected energy is small, due to the extremely short duration of the EMP.

The means by which the electromagnetic field of EMP may couple into the system are often referred to as the modes of entry of EMP energy. To predict the damaging potential of EMP to a system, one must investigate these various modes and the voltages and currents arising therefrom for a threatening incident EMP field. Usually, it is a difficult task to determine precisely where and how the EMP incident energy will couple into the system. However, some major modes of entry could be considered to form a basis for the EMP vulnerability prediction. They are:

- Antennas
- Penetration through shields and enclosures
- Slots and slits in enclosures, doors and windows
- Cables and transmission lines
- Utility lines
- Hybrid coupling modes

Figure 3.1 illustrates typical modes of entry of EMP energy into an electronic or weapon system.

For a given system and a specified EMP environment, one must consider at least the dominant mode or modes of entry of EMP for the vulnerability assessment. In most cases, such dominant modes can be determined by a physical inspection of the system and by examining the sensitivity of its subsystems or components.

The next step is to determine the voltages and currents, sometimes fields, and energy induced by EMP through the dominant mode or modes at the sensitive point or region in a system. Such a determination can be made by theoretical analyses or experiments or both. Because of the severe limitations of analytical tools, some idealization of the problems actually encountered in real life and associated approximations must often be made to obtain analytically the quantitative estimates of voltages and currents resulting from any mode of entry and for any specified EMP environment. Theoretical analyses of even the idealized coupling mechanisms, however, are useful to gain an insight into the problem and often to make the design decision. Of course, confidence in the results of idealized theoretical analyses can be improved significantly by experimental validation of such results.

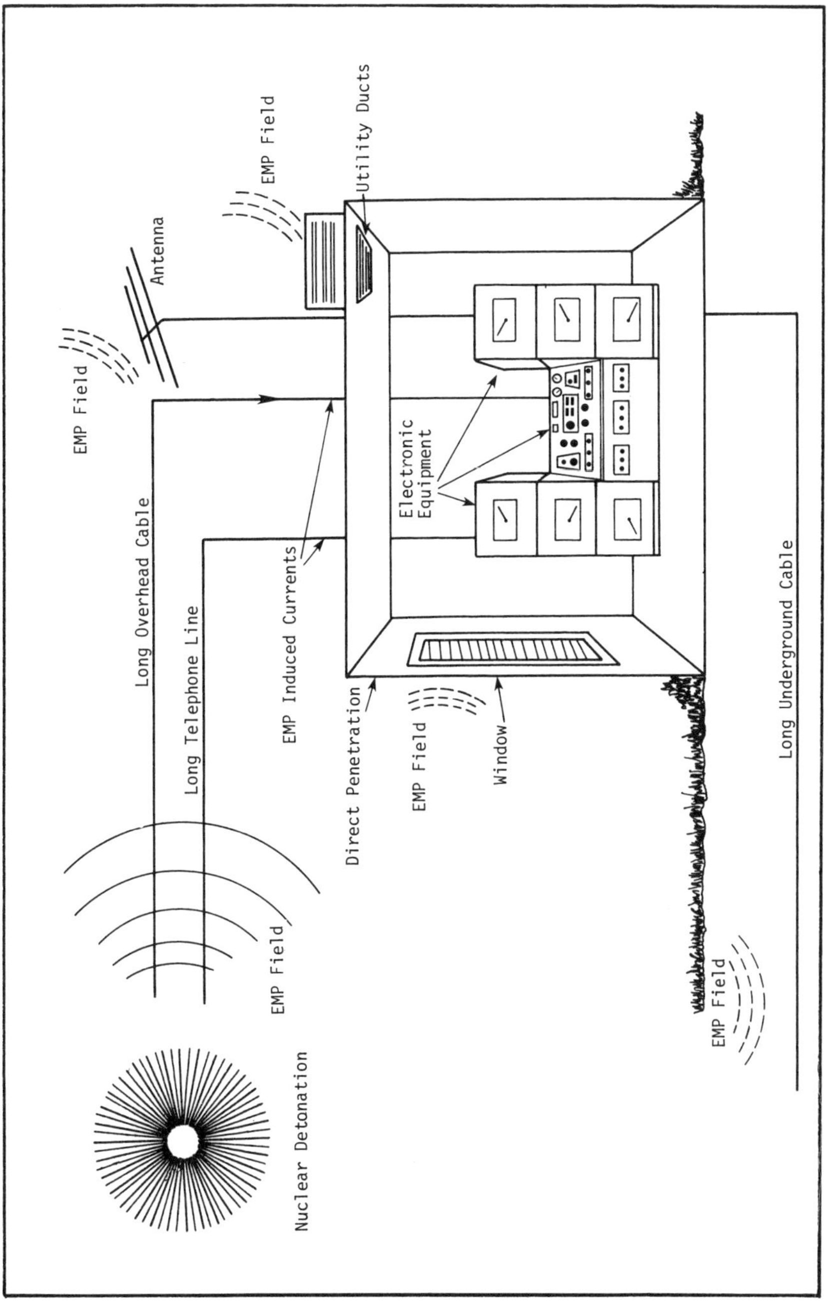

Figure 3.1 - Illustration of Modes of Entry of EMP Energy in an Electronic System.

Formulations of analytical problems relating to the various modes of entry and the approaches to their solutions are outlined in the following sections.

3.2 Coupling Through Linear Antennas and Arrays

In most cases, the antenna is perhaps the most efficient coupling mode of EMP energy into the electronic or weapon system, since it is designed to intercept as much energy from an impinging field as possible and to transfer this energy to the system as efficiently as possible. For the prediction of EMP vulnerability through the antenna coupling mode, one may first determine the voltage or current induced at the antenna feed-point for the specified impinging EMP field. For the purpose of such an analysis, it is convenient to classify antennas as:

- Linear Antennas
- Aperture Antennas
- Others

Linear antennas again can be further classified, depending on their uses, as:

- Dipole in free space
- Dipole over a ground plane
- Monopole over a ground plane
- Folded dipole
- Loop antenna
- Conical monopole
- Linear antenna arrays

The above list is far from being all-inclusive.

The analysis to determine EMP-induced voltage or current at the input of the linear antenna is somewhat different from the conventional antenna theory where the emphasis is on the type and the physical dimension of the antenna to optimize its performance at a fixed frequency or a frequency band. In the case of EMP, the type of antenna and its dimension are fixed, and the impinging field covers a very broad band of frequencies.

3.2.1 Monopole or Dipole Antenna Response Due to EMP

The transient response of a monopole or dipole antenna to the impinging EMP field appears to be that of a superposition of damped oscillations. The frequency of these oscillations is determined by the first resonant frequency of the antenna and the low-frequency response of the antenna. This is illustrated in Figure 3.1a.

The physical interpretation of the transient oscillations in terms of charges and currents can be given by considering the unloaded equivalent dipole antenna. Initially, before the arrival of EMP, the positive and negative antenna charges are equally distributed. After the arrival of EMP, and for the time less than that required for an electromagnetic wave to travel from the center to the end of the dipole, that is, the distance h as shown in Figure 3.1a, positive charges travel toward the antenna tip in one arm and negative charges in the other arm. The reflections of moving charges take place at the ends of

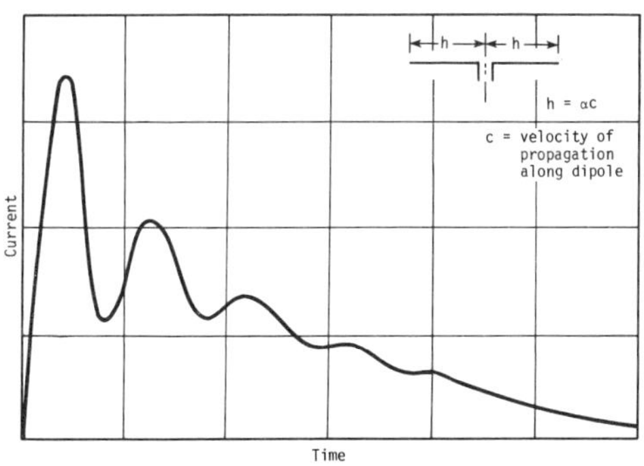

Figure 3.1a - Transient Current Response of a Monopole Antenna with a Resistive Load in Free-Space.

the dipole at the time $t = \alpha$ where, by analogy of the open-ended transmission line, there is a positive reflection coefficient. The reflected charges then move to the other end of the antenna and are reflected again from the open end at $t = 3\alpha$. This process repeats at $t = 4\alpha$. The period of oscillation is therefore determined by the time required for the charges to travel a distance four times the length of the monopole antenna. The gradual decay of the transient oscillations for an unloaded monopole as shown in Figure 3.1a is due to radiation losses.

To determine the antenna current, one may consider the equivalent circuit shown in Figure 3.2b. If the EMP incident or impinging field on the antenna is E (ω) and if the effective height of the antenna is $2h_e$, the open-circuit voltage induced at the monopole due to this field will be:

$$V_o = -2E^i(\omega)\ h_e\ (\omega, \theta) \tag{3.1}$$

θ being the angle of arrival of the EMP field. It should be remarked that E^i should be a function of angular frequency ω of the EMP field and h_e is a function of both ω and θ.

The current at the antenna load I_L is given by:

$$I_L = \frac{-2h_e E^i(\omega)}{Z_a + Z_L} \tag{3.2}$$

Z_a and Z_L being the antenna impedance and load impedance, respectively.

Schelkunoff* has developed expressions for voltages and currents for a thin-wire antenna by extending the theory of cone-transmission line and considering the propagation of transverse electromagnetic waves along the line. The expressions for the voltage and current as a function of distance r along the antenna from the center are

$$V(r) = V(o)\left[\text{Cos}\,\beta r + \frac{M(r)}{K_o}\text{Cos}\,\beta r - \frac{N(r)}{K_o}\text{Sin}\,\beta r\right] \tag{3.3}$$
$$-iK_o\, I(o)\left[\text{Sin}\,\beta r - \frac{M(r)}{K_o}\text{Sin}\,\beta r - \frac{N(r)}{K_o}\text{Cos}\,\beta r\right]$$

$$I(r) = \frac{-iV(o)}{K_o}\left[\text{Sin}\,\beta r + \frac{M(r)}{K_o}\text{Sin}\,\beta r + \frac{N(r)}{K_o}\text{Cos}\,\beta r\right] \tag{3.4}$$
$$+ I(o)\left[\text{Cos}\,\beta r - \frac{M(r)}{K_o}\text{Cos}\,\beta r + \frac{N(r)}{K_o}\text{Sin}\,\beta r\right]$$

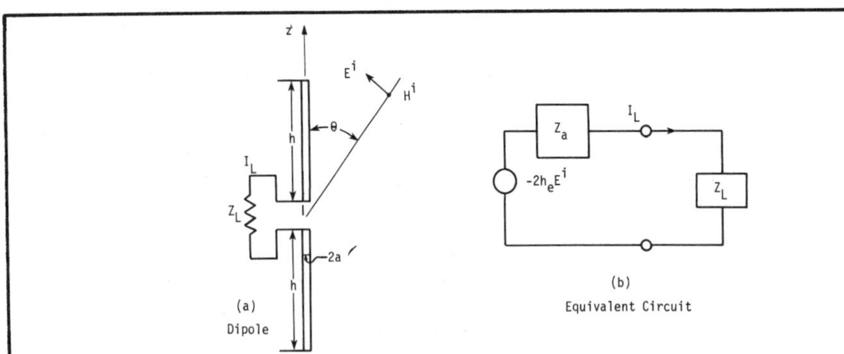

Figure 3.2 - Dipole Antenna and its Equivalent Circuit.

* S. A. Schelkunoff, "Electromagnetic Waves," D. Van Nostrand Co., Inc., New York, 1943, p. 291.

where $\beta = 2\pi/\lambda$, λ being the wavelength:

$$K_o = 120 \ln \frac{2h}{a} - 120$$

$$M(r) = 60 \, (1 - \cos 2\beta r)\left(\ln \frac{h}{r} - 1\right) + 60 \, (\ln 2\beta r - \mathrm{Ci}\,(2\beta r) + C)$$

$$N(r) = 60 \left(\ln \frac{h}{r} - 1\right) \sin \beta r + 60 \, \mathrm{Si}\,(2\beta r)$$

Ci x and Si x being, respectively, defined as the cosine integral and sine integral of x, C = 0.577, and a is the radius of the cylindrical antenna.

If the height or length of each arm of the dipole is h, one may observe from Equation 3.4 that:

$$I(h) = 0 \tag{3.5}$$

and
$$i \frac{V(o)}{I(o)} = K_o \left\{ \cos \beta h - \frac{M(h)}{K_o} \sin \beta h + \frac{N(h)}{K_o} \sin \beta h \right\} \div \left\{ \sin \beta h + \frac{M(h)}{K_o} \sin \beta h + \frac{N(h)}{K_o} \cos \beta h \right\} \tag{3.6}$$

When:

$\beta h << 1$, $\cos \beta h \simeq 1$, $\sin \beta h \simeq \beta h$

$\mathrm{Ci}\,(2\beta h) \simeq \ln 2\beta h + C$, $\mathrm{Si}\,(2\beta h) \simeq 2\beta h$

$M(h) \simeq 0$, $N(h) \simeq 0$

Therefore,

$$i \frac{V(o)}{I(o)} \simeq K_o \cot \beta h \simeq \frac{K_o}{\beta h} \tag{3.7}$$

Since $\frac{V(o)}{I(o)}$ denotes the input impedance of the antenna, it is evident that the antenna impedance becomes equivalent to a capacitor C_a, such that:

$$Z_a = \frac{1}{i\omega C_a} = \frac{K_o}{i\beta h} = \frac{c K_o}{i\omega h} \tag{3.8}$$

c being the velocity of propagation of the electromagnetic waves.

Thus, from Equation 3.8, the antenna capacitance becomes:

$$C_a \simeq h/K_o c \tag{3.9}$$

The total antenna circuit impedance Z_T, which affects the antenna circuit current, in this case is the sum of the load impedance and the impedance due to C_a, that is

$$Z_T = Z_L + \frac{1}{i\omega C_a}$$

If now $Z_L \simeq R_L << \frac{1}{\omega C_a}$, corresponding to a low-impedance load,

$$I_L = -i2h^2 E^i \operatorname{Sin}\theta\, \omega/K_o c \tag{3.10}$$

where h_e in Equation 3.1 is replaced by h Sin θ (Figure 3.1a).

If $E_i(t)$, the time-domain response of $E_i(\omega)$, is denoted by:

$$E^i(\omega) = \int_{-\infty}^{\infty} E^i(t)\, e^{i\omega t}\, dt$$

$$I_L(t) \simeq \frac{2h^2 \operatorname{Sin}\theta}{K_o c} \left| \frac{d}{dt} E^i(t) \right| \tag{3.11}$$

Equation 3.11 shows the relation of the current induced in the antenna circuit for an arbitrary incident EMP field $E^i(t)$ impinging on the antenna. The susceptibility of the system due to the antenna, if any, can now be examined from the estimate of the time-varying current.

When the actual antenna load is connected to the antenna through a transmission line, the impedance Z_A as seen looking into this line from the antenna end is:

$$Z_A(\omega) = Z_o \frac{Z_L + Z_o \tanh(\gamma \ell)}{Z_o + Z_L \tanh(\gamma \ell)} \tag{3.12}$$

where Z_L = terminating impedance of the transmission line

Z_o = characteristic impedance of the line

γ = complex propagation factor of the line
ℓ = length of the line

The voltage at the load end is given by:

$$V_L(\omega) = V_A(\omega)\left[\text{Cosh}\ \gamma\ell - \frac{Z_o}{Z_T}\ \text{Sinh}\ \gamma\ell\right] \qquad (3.13)$$

The equivalent circuit shown in Figure 3.2b is valid when the antenna is isolated in free space or far enough above the ground to ignore the ground-reflection effects. When the effect of ground cannot be ignored, the equivalent circuit needs to be modified.

3.2.2 EMP and Horizontal Dipole Above a Ground Plane

The schematic arrangement of a horizontal electric dipole above a ground plane and its equivalent circuit are shown in Figures 3.3a and b. The equivalent circuit denotes two equivalent sources due to the dipole and its image, the image source being proportional to $\Gamma(\omega)$, that is, the reflection coefficient of the ground. The reflection coefficient is not a function of the antenna but depends on the angle of arrival of the impinging field and the ground conductivity σ and dielectric constant ε_r. Thus, for horizontal and vertical polarizations, respectively:

$$\Gamma_H = \frac{\text{Sin}\ \Delta - \sqrt{M - \text{Cos}^2\ \Delta}}{\text{Sin}\ \Delta + \sqrt{M - \text{Cos}^2\ \Delta}} \qquad (3.14)$$

$$\Gamma_V = \frac{M\ \text{Sin}\ \Delta - \sqrt{M - \text{Cos}^2\ \Delta}}{M\ \text{Sin}\ \Delta + \sqrt{M - \text{Cos}^2\ \Delta}} \qquad (3.15)$$

where,
Δ = elevation angle
M = $\varepsilon_r - ix$
ε_r = relative dielectric constant of the ground
x = $\sigma / \omega\varepsilon_o$
σ = ground conductivity
ε_o = dielectric constant of free-space

The current in the antenna circuit can be written as:

$$I_L(\omega) = \frac{-2h_e E^i(\omega) - 2h_e \Gamma e^{-i\omega T} E^i(\omega)}{Z_a'(\omega) + Z_T(\omega)} \tag{3.16}$$

ωT being the additional time delay for the ground-reflected signal.

For numerical calculations, Z_a' may be assumed to be the same as Z_a discussed above, particularly when the antenna is far enough removed from the ground.

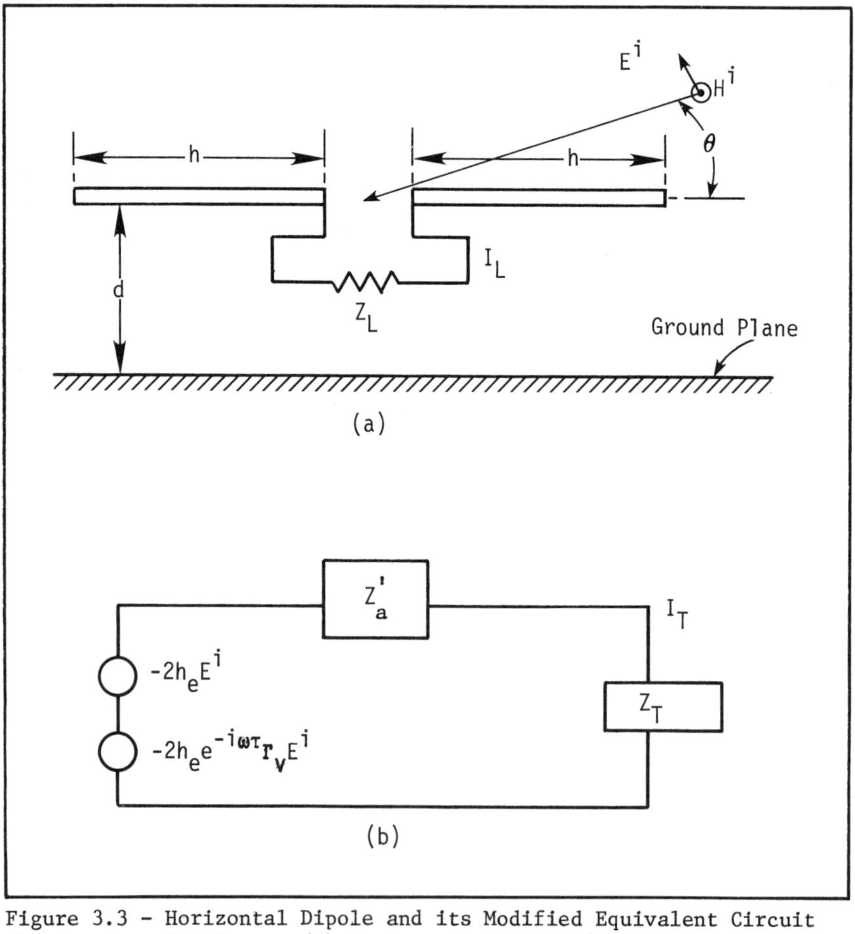

Figure 3.3 - Horizontal Dipole and its Modified Equivalent Circuit Showing Ground Effects. (a) Horizontal Dipole, (b) Equivalent Circuit.

3.2.3 EMP and Monopole Over a Ground Plane

Figure 3.4 shows the configuration of a vertical electric monopole antenna over a ground plane. The equivalent circuit for this case will be very similar to that shown in Figure 3.3b except that the factor $2h_e$ is replaced by h_e in the case of a monopole.

The time-domain response for the antenna current I_L as given in Equation 3.16 for the dipole case or as modified for monopole, is obtained from the inverse-Fourier transform. Thus,

$$I_L(t) = \frac{1}{2\pi} \int_{-\infty}^{\infty} I_L(\omega)\, e^{i\omega t} d\omega \tag{3.17}$$

To obtain $E^i(\omega)$ in Equation 3.16 when considering the time-varying nature of the impinging EMP field, one also needs to employ the Fourier transform. Thus,

$$E^i(\omega) = \int_{-\infty}^{\infty} E^i(t)\, e^{-i\omega t} dt \tag{3.18}$$

For numerical computations, one may approximate $E^i(t)$ if needed to obtain a pulse shape which is readily integratable.

The use of the Fourier transform to obtain the time-varying antenna current for a given incident time-varying EMP field $E^i(t)$ is convenient and straightforward as long as the load impedance is linear. For a non-linear load, computer programs may be developed to obtain the time-domain current response directly from $E^i(t)$.

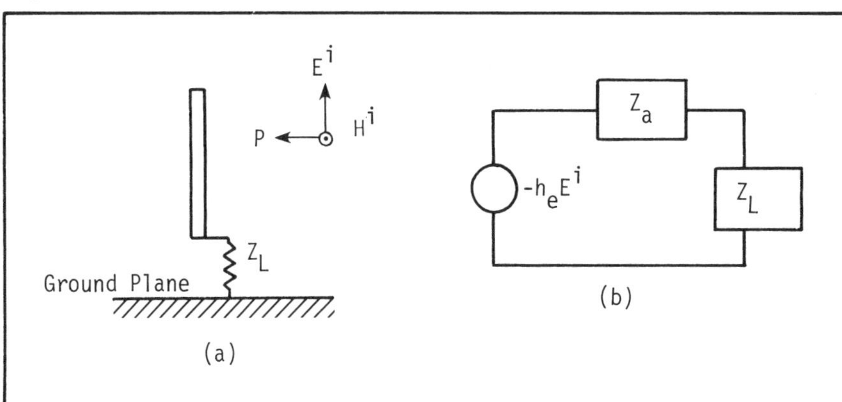

Figure 3.4 - Electrical Monopole Over a Ground Plane. (a) Monopole Antenna, (b) Equivalent Circuit Without Showing Ground Effects. Z_a is Antenna Impedance, Z_L is Antenna Load Impedance.

3.2.4 EMP and Vertical Monopole on a Conducting Mast

Often a monopole receiving antenna is placed on a mast. Figure 3.5 shows such an arrangement. This antenna is equivalent to asymmetrically-driven dipole antennas. Thus, for example, the input impedance of the asymmetrically-driven monopole is approximately given by:

$$Z_a(\omega) = Z_{ah}(\omega) + Z_{A(2H+h)}(\omega) \tag{3.19}$$

where: $Z_{ah}(\omega)$ = impedance of a monopole of length h and radius a above an infinite ground plane

$Z_{A(2H+h)}(\omega)$ = impedance of a monopole of length 2H+h and radius A above an infinite ground plane

The effective length of the asymmetrically-driven monopole is approximately given by:

$$h_e(\omega) \simeq h_{eh}(\omega) + h_{eH}(\omega) \tag{3.20}$$

where: $h_{eh}(\omega)$ = effective height of the monopole antenna of height h

$h_{eH}(\omega)$ = effective height of the monopole antenna of height H.

Once Z_a and h_e of the antenna are known, one can obtain a frequency-domain response for the antenna circuit current similar to Equation 3.2 for any known impinging field $E^i(\omega)$. Inverse-Fourier transform technique, as discussed earlier, can then be used to determine the time-varying antenna current for a time-varying $E^i(t)$ due to EMP.

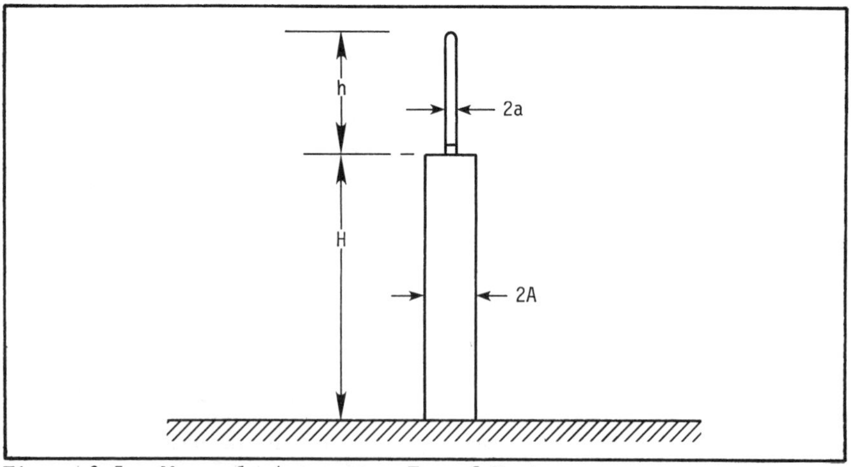

Figure 3.5 - Monopole Antenna on Top of Mast.

3.2.5 EMP and Folded Monopole

Sometimes, a folded monopole above a ground plane is used as a receiving antenna. EMP energy coupling into the receiving system through such an antenna can be determined from the antenna circuit current in this case. Figure 3.6a shows such an antenna arrangement. The corresponding antenna circuit current is given by:

$$I_L(\omega) = - \frac{4h_e \; Z_f \; E^i(\omega)}{(Z + Z_L) \; (Z_f + 2Z_r)} \qquad (3.21)$$

where:

h = effective height of an equivalent monopole of height h and radius $a_e = \sqrt{as}$

a = radius of the individual conductors that make up the monopole

s = separation distance between the axis of the wires as shown in the figure

Z_f = $iZ_o \tan \omega h/c$

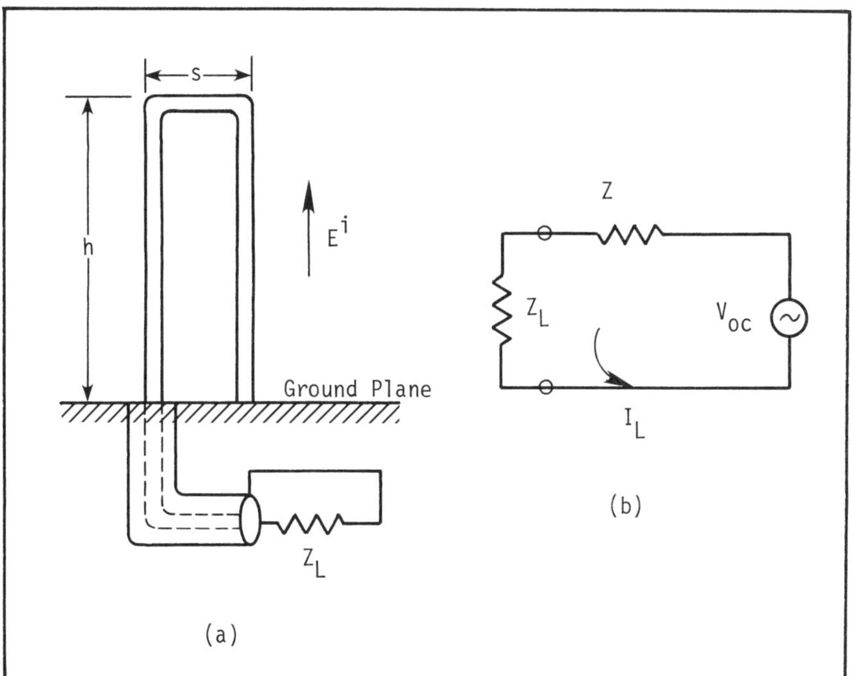

Figure 3.6 - Folded Monopole and its Equivalent Circuit. (a) Folded Monopole over Ground Plane, (b) Equivalent Circuit of Monopole.

Z_o = characteristic impedance of a two-wire transmission line composed of two wires of radius a and separation distance s

c = velocity of propagation

$$Z = \frac{4\, Z_r\, Z_f}{2\, Z_r + Z_f}$$

The equivalent circuit for the antenna is shown in Figure 3.6b. Again, the time-domain response is obtained from the inverse-Fourier transform of Equation 3.20.

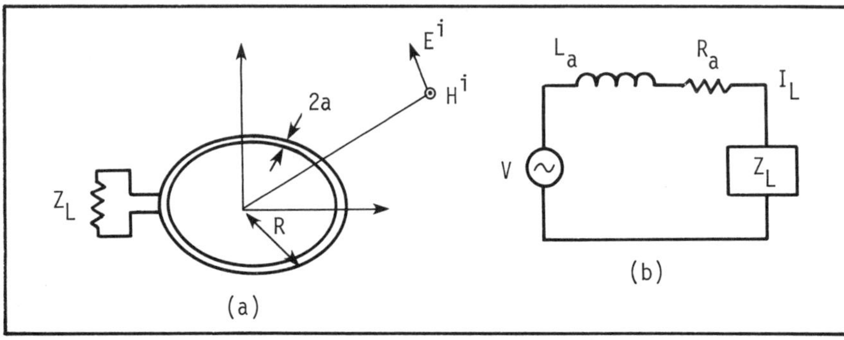

Figure 3.7 - Magnetic Dipole or Loop Antenna and its Equivalent Circuit (a) Loop Antenna, (b) Equivalent Circuit.

3.2.6 Response of Loop Antenna Due to EMP

If the antenna is now in the form of a loop as shown in Figure 3.7a, the open circuit voltage, V_O , developed at the antenna by an arbitrary incident EMP field, can be expressed from Maxwell's equation as:

$$V_o = \oint \vec{E} \cdot \vec{ds} = \iint \dot{H}\ \mu_o\ \mathrm{Sin}\ \theta\, dS \tag{3.22}$$

where:

E = electric field at the segment, ds, along the loop
H = incident magnetic field of EMP
μ_o = free-space permeability
θ = angle between the direction of arrival and the loop-normal
$\dot{H} = \frac{dH}{dt}$

The integration at the right side in Equation 3.22 is performed over the area of the loop.

The equivalent antenna circuit with load is shown in Figure 3.7b. The parameters L_a and R_a, respectively, denote the inductance and the resistance of the loop antenna. The resistive term R_a includes the radiation resistance and the ohmic resistance, i.e., the resistance of the loop antenna wire at the frequency of concern. The load impedance Z_L along with the resistance and inductance of the antenna determine the current in the antenna circuit for any given open-circuit voltage, V.

For a circular loop, the inductance of a loop antenna is given by:

$$L_a = R \mu_o \left[\ell n \left(\frac{8R}{a}\right) - 2 \right] \quad \text{for } R >> a \qquad (3.23)$$

where:
- R = radius of the loop in meter
- a = radius of the antenna wire in meter
- μ_o = free-space permeability

The corresponding inductance for a rectangular loop antenna, with sides ℓ_1 and ℓ_2 in meter, is given by:

$$L_a = \frac{\mu_o}{\pi} \left[\ell_1 \operatorname{Cosh}^{-1} \left(\frac{\ell_2}{2a}\right) + \ell_2 \operatorname{Cosh}^{-1} \left(\frac{\ell_1}{2a}\right) \right] \qquad (3.24)$$

The radiation resistance of a loop antenna can be determined from the power radiated by the loop antenna per unit current. From Schelkunoff*, the radiated power in free-space from a uniform current loop of any radius, R, is:

$$W = 30 \pi^2 \beta R \, I^2 \int_0^{2 \beta R} J_2(x) \, dx \text{ watts} \qquad (3.25)$$

where:
- β = 2 π/wavelength
- I = current in amperes
- $J_2(x)$ = Bessel function of second order of x

Since the radiation resistance, R_r, is defined as:

$$W = \frac{1}{2} I^2 R_r, \qquad (3.26)$$

---

* opt. cit., p. 509.

from Equations 3.25 and 3.26, R_r can be expressed as:

$$R_r = 60\,\pi^2\,\beta R \int_o^{2\beta R} J_2(x)\,dx \tag{3.27}$$

When, $2\beta a << 1,$

$$J_2(x) \approx \frac{x^2}{2^2\;2!}$$

$$\int_o^{2\beta R} J_2(x)\,dx = \frac{\beta^3 R^3}{3} \tag{3.28}$$

and, $R_r = 20\,(\beta^2 S)^2$

where S is the area of the loop.

It may be noted that the radiation resistance of a small loop as assumed in Equation 3.28 remains approximately the same as given in Equation 3.28, regardless of the geometry of the loop.

The ohmic resistance of the loop antenna is:

$$R_{ohmic} = 2\pi R\rho \quad \text{for circular loop}$$

$$= 2\,(\ell_1 + \ell_2)\rho \quad \text{for rectangular loop} \tag{3.29}$$

ρ being the unit length resistance of the loop-antenna wire in ohms/meter.

Let: $$H(\omega) = \int_{-\infty}^{\infty} H(t)e^{-i\omega t}\,dt \tag{3.30}$$

where H(t) is the time-varying incident magnetic field due to EMP. Then, from Equation 3.22:

$$V(\omega) = \iint i\;H(\omega)\;\mu_o\;\text{Sin}\,\theta\;dS \simeq i\omega H(\omega)\;\mu_o\;\text{Sin}\,\theta\;S \tag{3.31}$$

for a small loop, S being the area of the loop.

The current in the loop antenna can now be written as:

$$I_L(\omega) = \frac{V(\omega)}{R_r + R_{ohmic} + i\omega L_a + Z_L} \tag{3.32}$$

where Z_L is the load impedance as shown in Figure 3.7b.

For high-impedance load, i.e., for:

$$|Z_L| >> |R_r + R_{ohmic} + i\omega L_a| , \quad I_L(\omega) = \frac{V(\omega)}{Z_L} \tag{3.33}$$

and voltage across the antenna load, $V_L = V(\omega)$.

Similarly, for low-impedance resistive load,

$$Z_L = R_L << \omega L_a$$

$$I_L(\omega) = \frac{V(\omega)}{R_r + R_{ohmic} + i\omega L} \tag{3.34}$$

For both Equations 3.33 and 3.34,

$$I_L(t) = \frac{1}{2\pi} \int_{-\infty}^{\infty} I_L(\omega)\, e^{i\omega t} d\omega \tag{3.35}$$

Thus, one can determine the actual time-varying current in the antenna circuit for any arbitrary, but given, time-varying incident H field due to EMP.

3.2.7 EMP and Conical Monopole

A conical monopole is used in a receiving system when a broadband signal reception is required uniformly from all directions normal to the antenna axis. Figures 3.8a and b show the antenna arrangement and its equivalent circuit. The effective antenna height in this case is:

$$h_e = \frac{6\pi a Z_o \operatorname{Cos} \theta_o}{4\pi Z_o + 3\eta_o \operatorname{Cos}^2 \theta_o} \tag{3.36}$$

where $Z_o = 60 \ln \cot (\theta_o/2)$

θ_o = angle of the cone as shown in the figure

η_o = intrinsic impedance of free space = 377 ohms.

The antenna impedance, which is very nearly a capacitive reactance, is given by:

$$Z_a = \frac{1}{i\omega C_a} \tag{3.37}$$

where:
$$C_a = \frac{a\,(Z_o + 90 \cos^2 \theta_o)}{c\ Z_o 2}$$

The antenna circuit current then can be written as:

$$I_L(\omega) = \frac{-h_e E^i(\omega)}{\frac{1}{i\omega C_a} + Z_L} \tag{3.38}$$

Again, the time-domain expression for I_L can be obtained from the inverse-Fourier transform, provided, of course, the antenna load is linear.

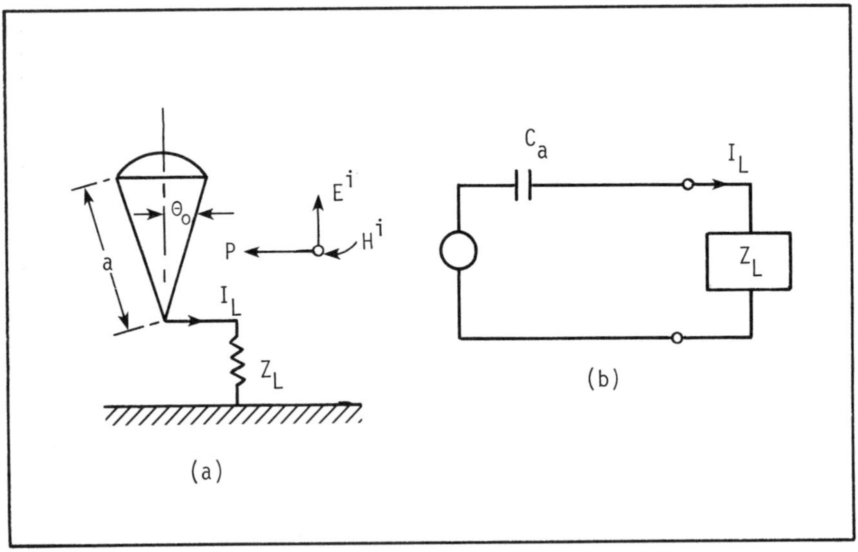

Figure 3.8 - Conical Monopole and its Equivalent Circuit. (a) Conical Monopole over Ground Plane, (b) Equivalent Circuit of Conical Monopole.

3.2.8 Response of Linear Arrays Due to EMP

In addition to the linear antennas discussed above, sometimes an array of such antennas is used in a receiving system. Discussions and analyses for all possible configurations of linear antenna arrays are beyond the scope of this handbook. However, an illustration of a two-element array may outline an approach for such a case. Figure 3.9 shows an array of two parallel linear dipoles driven in phase for a broadside antenna pattern or in phase-opposition for an end-fire pattern.

The antenna circuit current for such an array is given by:

$$I_L = \frac{-2\,h_e\ E^i(\omega)\ A\ (\phi)\ K_1}{Z_L + Z_{in}} \qquad (3.39)$$

where:

$E^i(\omega)$ = incident E field of EMP as a function of ω

$2\,h_e$ = effective height of each dipole

$$Z_{in} = \frac{Z_1\ Z_{al}}{Z_1\ Z_{al}}$$

$$Z_1 = \frac{G\ D\ Z_o}{Z_o(Z_{sl}\ F - Z_{12}\ G)\ \mathrm{Cos}\ \theta + i\,FD\ \mathrm{Sin}\ \theta}$$

$$Z_{al} = \frac{GD}{Z_{sl}\ G - Z_{12}\ F}$$

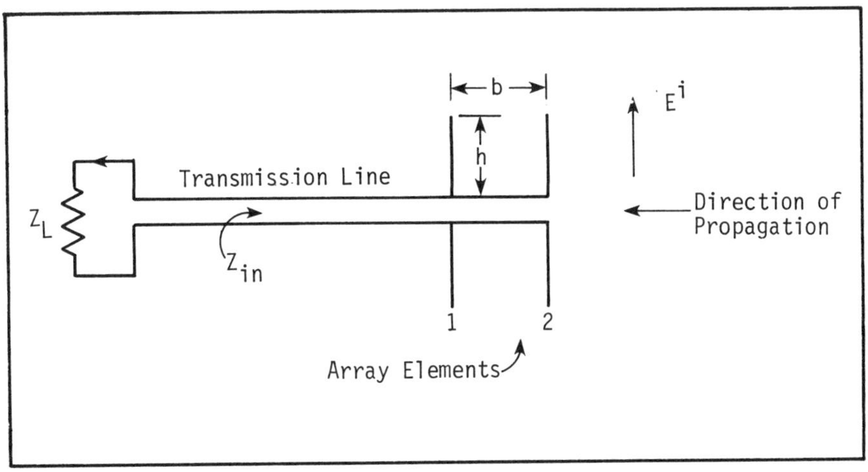

Figure 3.9 - Two Element Receiving Antenna.

$\theta = \omega b/c$

c = velocity of propagation

$D = Z_{s1}^{2} - Z_{12}^{2}$

$F = D + i\, Z_o\, Z_{12} \operatorname{Sin}\theta$

$G = D \operatorname{Cos}\theta + i\, Z_o\, Z_{s1} \operatorname{Sin}\theta$

Z_o = characteristic impedance of the transmission line connecting the dipoles

$$K_1 = \frac{Z_1}{Z_1 + Za2}$$

$$Z_{a2} = \frac{FD}{Z_{s1}\, F - Z_{12}\, G}$$

Z_{s1} = self-impedance of two parallel antennas

Z_{12} = mutual impedance of two parallel antennas

$A(\phi) = 1 + K_2\, e^{ib \operatorname{Cos}\phi}$

$$K_2 = \frac{Z_{a1}}{Z_{a2} \operatorname{Cos}\theta + iZ_o \operatorname{Sin}\theta}$$

b = distance between antenna elements

ϕ = azimuthal angle measured from the plane of the array.

The self and mutual impedances, Z_{s1} and Z_{12}, referred to above as a function of ($\omega h/c$) are shown in Figures 3.10 and 3.11.

If the distance between the elements of the array is set equal to a half wavelength at the center frequency, f_o, the antenna will operate as an end-fire, two-element array at f_o with the maximum reception at $\phi = 0$. If the direction of propagation of the incident field $E^i(\omega)$ is along the $\phi = 0$ plane, the array factor becomes:

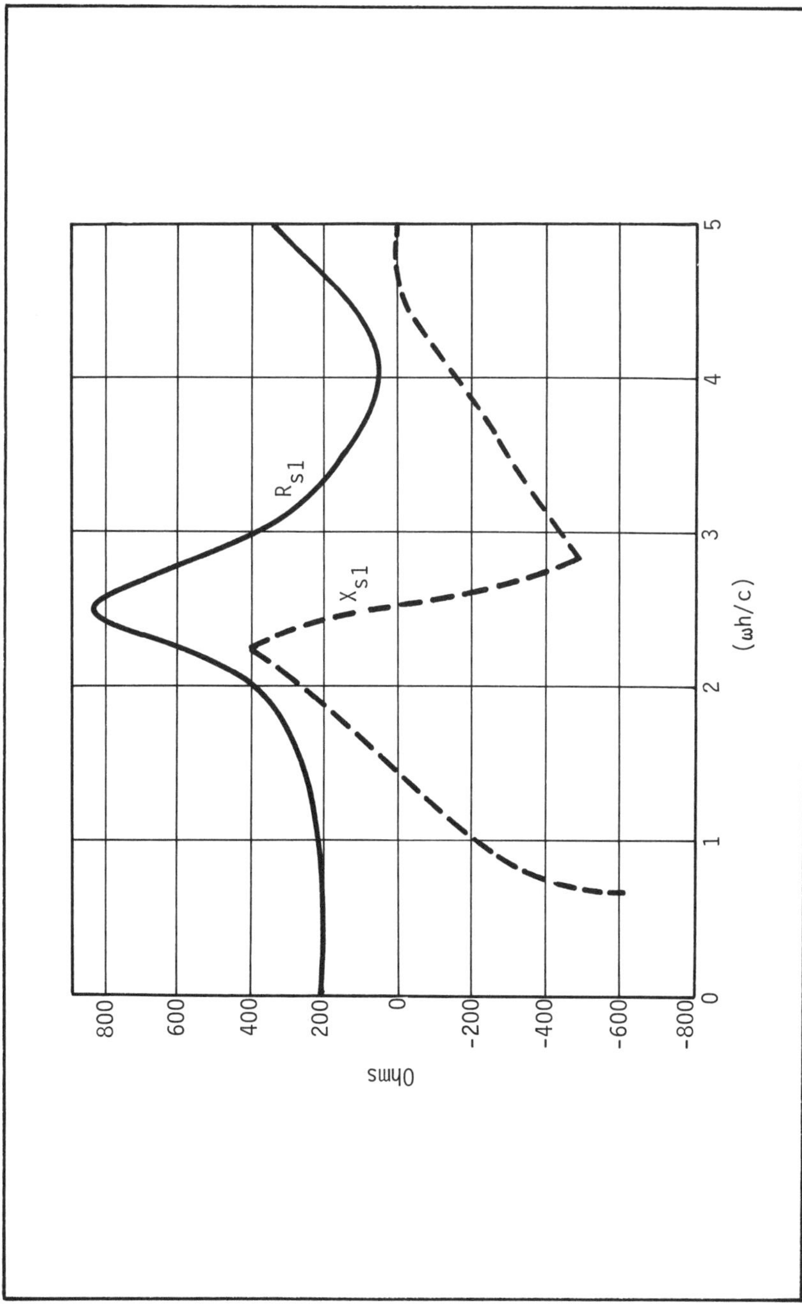

Figure 3.10 - Self-impedance of Two Parellel Dipole Antennas in an Array. R_{s1} is the Real Part of the Impedance. X_{s1} is the Imaginary part of the Impedance.

$$A = 1 + K_2 e^{i\beta b}$$

The analyses for other linear arrays are straightforward. The expression for the antenna circuit current $I_L(\omega)$ becomes more involved. The corresponding expression $I_L(t)$ in the time domain, which can be obtained from the inverse-Fourier transform of $I_L(\omega)$ also becomes involved and a computer aid may be necessary.

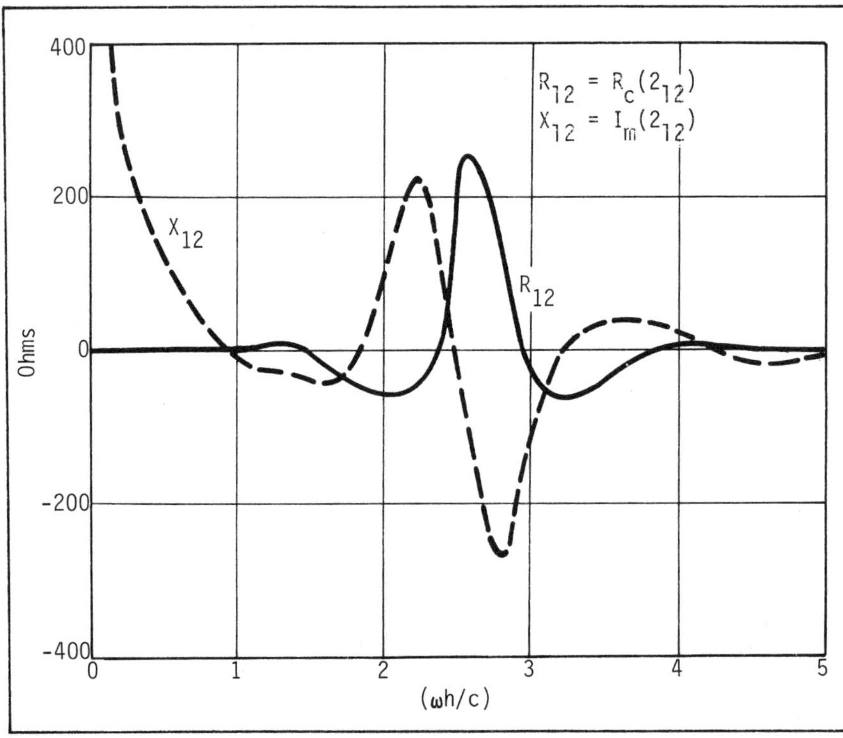

Figure 3.11 - Mutual Impedance of Two Parallel Dipole Antennas In An Array.

R_{12} Is the Real Part of the Impedance.

X_{12} Is the Imaginary Part of the Impedance.

3.3 Coupling Through Aperture Antennas

Aperture antennas such as microwave horns and dishes are frequently used for radar, communication and navigational receivers. It is desirable to assess the susceptibility of these receivers when the antenna couples electromagnetic energy of EMP into the system.

Although the essential analytical steps necessary to calculate electromagnetic energy injection into the receiver are the same as those used for linear antennas, there are some important differences. For example, in a linear antenna, the open-circuit voltage across the antenna is obtained by multiplying the effective height of the antenna with the incident electric field of EMP, $E^i(\omega)$, the effective height being a function of the polarization of the antenna with respect to that of $E^i(\omega)$. For aperture antennas, both linear or circular polarizations may be used, and conceptually it is difficult to find an effective height. Instead, one can determine the EMP power intercepted by the aperture antenna, and then determine the antenna circuit voltage or current from this power and the antenna circuit impedance. Secondly, aperture antennas are used at very high or microwave frequencies. Hence, the transfer function of the cable or waveguide connecting the antenna with the receiver must be taken into account to determine the power coupled into the receiver. A waveguide, however, has a cut-off frequency, and not all EMP power reaches the receiver regardless of the wide-frequency spectrum of EMP. Even when a waveguide is not used, the antenna connecting cable may have nonuniform frequency-dependent characteristics at frequencies other than those for which the antenna is designed. Since the time-varying voltage or current in an aperture antenna circuit resulting from EMP is obtained by performing inverse-Fourier transform of the EMP-intercepted power at all frequencies, it is necessary to determine the transfer function over the entire frequency band.

The ideal way to obtain the transfer characteristics of the aperture antenna circuit is to measure the amplitude and phase of the impedance (i.e. the complex impedance Z_A) as a function of frequency. Once Z_A is known, the antenna circuit current I_L can be determined from the relation,

$$I_L^2 (Z_A + Z_L) = P \tag{3.40}$$

where Z_L and P are respectively the antenna load impedance and the intercepted complex power. P is obtained by integrating the Poynting vector over the antenna aperture. Thus,

$$P = \iint E^i(\omega) \times H^i(\omega)\, dS \tag{3.41}$$

where: $E^i(\omega)$ = electric field due to EMP

$H^i(\omega)$ = magnetic field due to EMP

and the integration is over the aperture. P can also be written as:

$$E^i(\omega) \times H^i(\omega) \cdot A$$

when A is the effective area of the aperture antenna.

3.3.1 Aperture Antenna Response Due to EMP

In many cases, the exact transfer characteristics of an aperture antenna are not known and cannot be easily measured. An estimate of real EMP power coupled into the receiving system can be formulated with the help of an equivalent antenna circuit. Since most antenna circuits are designed to reject intercepted energy by the aperture beyond a narrow passband, the equivalent impedance of the antenna may be approximated* by one consisting of a tuned circuit with R-L-C elements as shown in Figure 3.12. The generated voltage in this figure is related to the instantaneous incident or intercepted power P of EMP, where

$$P = \vec{E} \times \vec{H}^* \cos\theta\, A \qquad (3.42)$$

$\vec{E}$ = incident electric field of EMP at the aperture

$\vec{H}^*$ = complex conjugate of the corresponding magnetic field of EMP at the aperture

θ = the angle between the normal to the aperture and the direction of propagation of the incident field.

A = effective area of the aperture antenna

The instantaneous power delivered by the generator in Figure 3.12 is

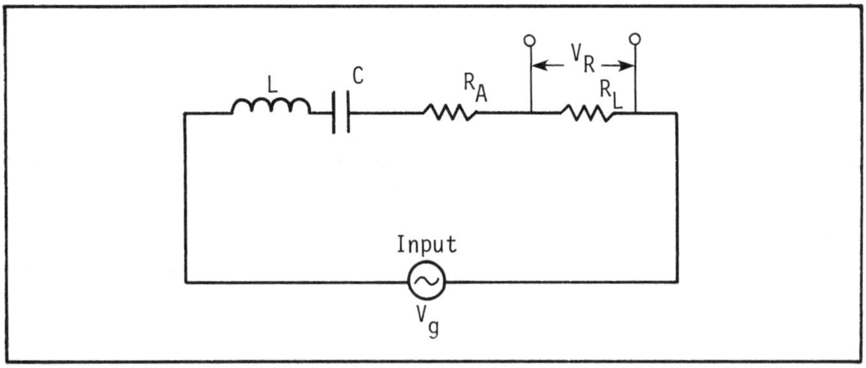

Figure 3.12 - An Approximate Equivalent Circuit for an Aperture Antenna and its Load.

* It may be noted that the effect of this approximation is to ignore the contribution of EMP outside the antenna-circuit passband.

$$P_g = \frac{V_g^{\,2}}{R_A + R_L} = P \qquad (3.43)$$

where: V_g = equivalent voltage due to EMP field

R_A = resistive part of the antenna impedance
R_L = resistive part of the load impedance

Thus,
$$V_g = \left[\vec{E} \times \vec{H}^* \operatorname{Cos} \theta \, A \, (R_A + R_L) \right]^{1/2} \qquad (3.44)$$

For a plane wave incident on the antenna:

$$|\vec{E} \times \vec{H}^*| = E^2/Z_o \qquad (3.45)$$

where Z_o is the characteristic impedance of the propagation medium, which is 377 ohms in free space.

When the antenna load impedance is a resistance R_L, the voltage across R_L can be expressed as:

$$V_L = \frac{E_o \, R_L/R_T}{1 - iQ\left(1 - \frac{\omega^2}{\omega_o^{\,2}}\right) \frac{\omega o}{\omega}} \qquad (3.46)$$

where: $E_o = E^i \sqrt{\frac{A \, R_T \, \operatorname{Cos} \theta}{Z_o}}$

$Q = \omega_o L/R_T, \quad \omega_o L = 1/\omega_o C$

and $R_L = R_A + R_L$

L and C being the equivalent inductance and capacitance of the antenna circuit as shown in Figure 3.12.

3.3.2 Time-Domain Response of Aperture Antenna

In Equation 3.46 E^i is a function of ω, that is, $E^i(\omega)$ is the Fourier transform of the incident electric field. The transient voltage across the load can therefore be expressed as

$$V_L(t) = \frac{1}{2\pi} \int_{-\infty}^{\infty} \frac{E_o R_L/R_T}{1 - iQ\left(1 - \frac{\omega^2}{\omega_o^2}\right)\frac{\omega_o}{\omega}} e^{i\omega t} d\omega \quad (3.47)$$

It is often convenient to determine the transient response across the antenna load for a delta-function source since the short-duration EMP resulting from a nuclear detonation can be approximated by such a source. Also, when the response due to a delta-function source is known, the transient response for an arbitrary source may be obtained from a convolution integral of the product of the delta-function response and the actual source function. For a delta-function source:

$$E_o = E \sqrt{\frac{A R_T \cos\theta}{Z_o}} = \text{a constant function of } \omega$$

and,

$$V_L(t) = \frac{E_o R_L}{R_T} \frac{1}{\sqrt{1 - 1/Q^2}} e^{-\beta t} \cos(\alpha t - \psi) \quad (3.48)$$

where,

$$\alpha = \sqrt{1 - 1/4Q^2}$$

$$\beta = 1/(2Q)$$

$$\psi = \sin^{-1}\left(\frac{1}{2Q}\right)$$

The delta function response of the antenna becomes a damped sinusoid where the exponential damping factor is inversely proportional to the antenna circuit Q and is directly proportional to $\omega_o t$.

When the antenna load impedance is not resistive, i.e., $Z_L = R_L + i\omega L_L$, the time-domain response of the voltage across the load for a delta function source becomes:

$$\overline{V}_L(t) = V_L(t) + \left[\frac{L_L}{R_L} \frac{d V_L(t)}{dt}\right] \quad (3.49)$$

where $V_L(t)$ is similar to that given in Equation 3.47, except that Q and ω_o are appropriately modified to account for L_L. Thus, for this case,

$$Q = \frac{\omega_o (L + L_L)}{R_T} \tag{3.50}$$

$$\omega_o^2 (L + L_L) C = 1.$$

Similarly when the antenna load contains a capacitive reactance C_L, the transient response of V_L (t) becomes:

$$\bar{\bar{V}}_L (t) = V_L (t) + \frac{1}{C_L R_L} \int_o^t V_L (t') dt' \tag{3.51}$$

where V_L (t) is similar to that given in Equation 3.47 except that:

$$Q = \omega_o L/R_T$$

$$\omega_o^2 L (C + C_L) = 1 \tag{3.52}$$

Thus the time-varying function $\bar{V}_L$ (t) or $\bar{\bar{V}}_L$ (t) for an arbitrary load can be determined from V_L (t) corresponding to a resistive load as shown in Equations 3.49 and 3.51.

3.4 Coupling Through Shields and Enclosures

The energy in EMP can couple into electronic systems or components inside an enclosure through direct penetration of the enclosure walls. To determine the field inside an enclosure one may first find the transfer function (CW response) of the enclosure*. The inverse-Fourier transform of the product of the transfer function and the Fourier transform of the incident time-varying field due to EMP, then, yields the time-varying field inside the enclosure wall.

The derivation of the transfer function for an arbitrary shape and size of the enclosure is involved, and because of the complexity of the boundary value problem, a rigorous solution for the transfer function has been obtained in the past for ideal geometrics such as a sphere, infinite cylinder, etc.

Although the type of enclosure encountered in real life very seldom resembles an infinite cylinder, the transfer function for an infinite cylinder of any arbitrary thickness and wall material provides an insight into the nature of the field penetration through a wall. A derivation of this transfer function should be appropriate here.

3.4.1 Analytical Estimation of EMP Coupling Through Cylindrical Shields

Let a plane wave due to EMP be considered incident on an infinitely long, hollow cylinder. The field inside the cylinder can be derived from a Hertz vector with a single z (axial) directional component. Thus the electromagnetic field inside the cylinder can, in effect, be derived by a scalar potential which satisfies the wave equation:

$$\nabla^2 \psi - \mu\varepsilon \left[\frac{\partial^2 \psi}{\partial t^2}\right] - \mu\sigma \frac{\partial \psi}{\partial t} = 0 \qquad (3.53)$$

such than: $$H_z = \beta^2 \psi, \quad E_\phi = i\omega\mu \left[\frac{\partial \psi}{\partial r}\right]$$

and, $$E_r = \frac{i\omega\mu}{r} \left[\frac{\partial \psi}{\partial \phi}\right], \quad \beta^2 = \omega^2 \mu\varepsilon$$

where μ, ε and σ are the permeability, dielectric constant and conductivity of the medium and r, ϕ and z are the coordinates of the cylindrical coordinate system.

---

* The transfer function in this case is the ratio of the field inside the enclosure to that of the CW incident field.

Figure 3.13 shows a cross section of the hollow cylinder with its inner radius b and outer radius a. Designating the regions $r \leq b$, $b \leq r \leq a$ and $r \geq a$ as regions 1, 2 and 3, one may write the solutions of Equation 3.53 for the three regions as:

$$\psi_1 = e^{i\omega t} \sum_{n=o}^{\infty} A_n J_n (\beta_1 r) \text{ Cos } n\phi$$

$$\psi_2 = e^{i\omega t} \left\{ \sum_{n=o}^{\infty} B_n J_n (\beta_2 r) \text{ Cos } n\phi + \sum_{n=o}^{\infty} C_n H^{(2)} (\beta_2 r) \text{ Cos } n\phi \right\}$$

$$\psi_3 = e^{i\omega t} \sum_{n=o}^{\infty} D_n H_n^{(2)} (\beta_1 r) \text{ Cos } n\phi + \psi_o \quad (3.54)$$

where,

ψ_o = potential of the incident electromagnetic field

$J_n(x)$ = Bessel function of x of order n

$H_n^{(2)}(x)$ = Hankel function of the second kind of x

$\beta_1^2 = \omega^2 \mu_o \varepsilon_o$

$\beta_2^2 = \omega^2 \mu_2 \varepsilon_2 - i\omega\mu_2 \sigma_2$

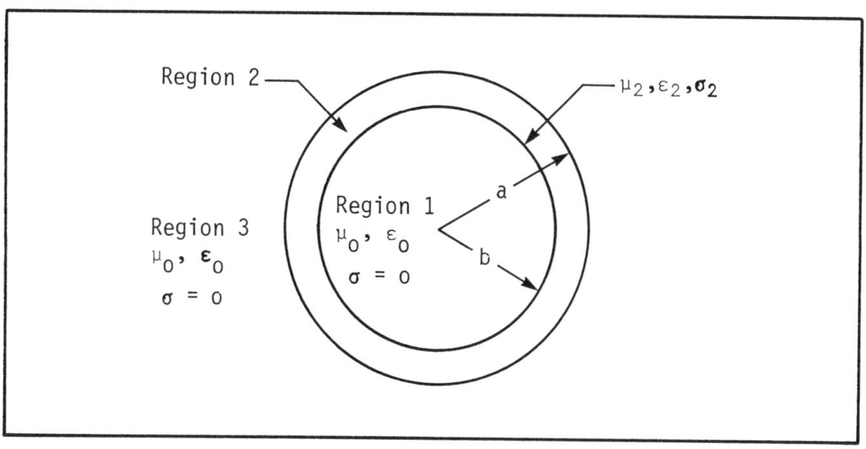

Figure 3.13 - Cross-section of a Hollow Infinite Cylinder.

μ_o and ε_o being respectively the permeability and dielectric constant of free space, and μ_2 , ε_2 and σ_2 being the permeability, dielectric constant and conductivity of the material forming the hollow cylinder.

For a plane incident electromagnetic field due to EMP,

$$\psi_o = \left[\frac{M}{\beta_1^{\,2}} J_o(\beta_1 r) + 2 \sum_{n=1}^{\infty} (-1)^n J_n(\beta_1 r) \operatorname{Cos} n\phi\right] \tag{3.55}$$

where M is the amplitude of the incident magnetic field.

The boundary conditions to be imposed on the solutions given in Equation 3.54 are that H_z and E_ϕ components of the field are continuous at each boundary. Equating term-wise for each n, one can obtain a formal solution of A_n from Equation 3.54 and the boundary conditions.

Although the formal solution should be used for the generalized case, one may obtain a simplified expression for the transfer function for the special cases where $\beta_1 a << 1$ and $\beta_2 a >> 1$. Many practical enclosures of concern fall into this group; therefore, a solution for this region is useful for practical problems. When $\beta_1 a << 1$, only $n = o$ term of the potential function becomes dominant, and under such a circumstance,

$$E_\phi = e^{i\omega t}\left[\frac{i\mu_1\ \omega\beta_1}{2} A_1 \operatorname{Cos}\phi\right]$$
$$H_z = e^{i\omega t}\left[\beta_1^{\,2} A_o\right] \tag{3.56}$$

Solving the equations arising from the boundary conditions as referred to above, and making use of the asymptotic expansion of Bessel and Hankel functions, corresponding to $|\beta_1 a| << 1$, and $|\beta_2 a| >> 1$, one obtains:

$$A_o = \frac{-2\,M\,(a/b)^{1/2}}{\beta_1^{\,2}\left\{\frac{\mu_1}{\mu_2}\beta_2 b \operatorname{Sin}\beta_2(a-b) - 2\operatorname{Cos}\beta_2(a-b)\right\}} \tag{3.57}$$

The transfer function for H_z field becomes:

$$T_{H_z} = \left[\frac{-2\,(a/b)^{1/2}}{\frac{\mu_1}{\mu_2}\beta_2 b \operatorname{Sin}\beta_2(a-b) - 2\operatorname{Cos}\beta_2(a-b)}\right] \tag{3.58}$$

The total electric field inside the hollow cylinder for such a case is given by:

$$E = E_r \operatorname{Sin}\phi + E_\phi \operatorname{Cos}\phi = E_y = e^{i\omega t} A_1 \frac{i\mu_1 \omega \beta_1}{2} \qquad (3.59)$$

Since the incident electric field is:

$$E_o \frac{\mu_1 \omega}{\beta_1} M \qquad (3.60)$$

the field inside the cylinder, in terms of the incident electric field is given by:

$$E = \left[\frac{2 E_o}{\frac{\beta_2 a}{(\beta_1 a)^2} \frac{\mu_1}{\mu_2} \operatorname{Sin}\beta_2 (a-b) + 2 \operatorname{Cos}\beta_2 (a-b)} \right] \qquad (3.61)$$

Furthermore, where:

$$\beta_2 a >> 1, \quad \left| \frac{\mu 1}{2\mu_2} \frac{\beta_2 a}{(\beta_1 a)^2} \right| >> 1$$

and,

$$E = \left[\frac{2 E_o}{\frac{\beta_2 a}{(\beta_1 a)^2} \frac{\mu_1}{\mu_2} \operatorname{Sin} \beta_2 (a-b)} \right] \qquad (3.62)$$

The transfer function for the electric field for an infinite cylinder and impressed axial magnetic field due to EMP is readily obtained from Equation 3.62 since, by definition, it is E/E_o.

3.4.2 Coupling Through Non-Cylindrical Shields

A comparison of corresponding transfer functions for geometrical structures other than the infinite circular cylinder may be in order here to illustrate the validity of such a transfer function for enclosure shapes which are encountered in real life. For example, the magnetic shielding* by infinite parallel conducting plates** is given by

---

* G.L. Brown and R. N. Ghose, "EMP Energy Coupling Studies," ANC 19R-3, American Nucleonics Corporation, 14 Jan 1966

** Heinrich Kaden, "Wirbelstrome und Schirmung in der Nachrichtentechnik," Springer-Verlag, 1959

$$T(\omega) = \frac{1}{\cos \beta_2 d - \frac{1}{2} \frac{\mu_1}{\mu_2} \beta_2 b \sin \beta_2 d} \tag{3.63}$$

where:
- T = ratio of the internal to incident field
- d = thickness of plates
- b = separation distance between the plates

Similarly, magnetic shielding by an infinite cylinder for an incident magnetic field perpendicular to the cylinder axis has been investigated by Brown and Ghose and Kaden. The transfer functions for such a case are

$$T_r = \left[\frac{\cos \theta}{\cos \beta_2 d - \frac{1}{2} (K - \frac{1}{K}) \sin \beta_2 d}\right] \tag{3.64}$$

$$T_\theta = \left[\frac{\sin \theta}{\cos \beta_2 d - \frac{1}{2} (K - \frac{1}{K}) \sin \beta_2 d}\right] \tag{3.65}$$

where:
- T_r = ratio of the radial component of the internal field to the incident field
- T_θ = ratio of the azimuthal component of the internal field to the incident field
- $K = \frac{\mu_1}{\mu_2} \beta_2 b$
- b = radius of cylinder

In this case, the transfer function for the absolute value of the magnetic field inside the cylinder is:

$$T = \left[\frac{1}{\cos \beta_2 d - \frac{1}{2} (K - \frac{1}{K}) \sin \beta_2 d}\right] \tag{3.66}$$

This expression is very similar to that given in Equation 3.58 where $(a - b) = d$ and $(a/b) \simeq 1$, and $K >> 1$ which is true for frequencies greater than a few Hertz. Thus the transfer function is independent of the polarization of the incident field.

In the case of a sphere, the magnetic shielding as obtained by Kaden* is:

$$T_{Sphere} = \frac{1}{\cos \beta_2 d - \frac{1}{3}(K - \frac{2}{K}) \sin \beta_2 d} \qquad (3.67)$$

A similar expression as obtained by King** for a non-magnetic sphere is:

$$T(\omega) = \left[\frac{1}{\cos \beta_2 d - \frac{1}{3}(K - \frac{1}{K}) \sin \beta_2 d} \right] \qquad (3.68)$$

where, $K = \frac{\mu_1}{\mu_2} \beta_2 b$

b = radius of the sphere

From Equations 3.63, 3.66 and 3.68, it is seen that the expressions for the transfer functions for parallel plates, infinite cylinder and a sphere are almost the same, the difference being that for a parallel plate, the parameter b is the plate separation, while for an infinite cylinder, b is equal to its radius, and for a sphere, b is equal to 2/3 the radius of the sphere. The fact that the transfer function is very nearly the same for these various shapes of the enclosure implies that the transfer function is perhaps not very much different for shapes of enclosures other than those considered above. More specifically, the transfer function for a rectangular enclosure is likely to be very similar to that for an infinite cylinder, at least near the center, although the field concentrations near the edges need to be taken into account for a rectangular enclosure.

3.4.3 Time-Varying Field Inside a Shielded Structure

Once the transfer function (as a function of ω) is known, the EMP field penetrating through the enclosure wall can be obtained by multiplying the transfer function with $E^i(\omega)$, or $H^i(\omega)$, i.e., the incident electric or magnetic field due to EMP. Also, as discussed before, the time-domain response of the field inside the enclosure is obtained from the inverse-Fourier transform of the multiplication product $E^i(\omega)T_E(\omega)$ or $H^i(\omega)T_H(\omega)$. An alternate method

---

* opt. cit

** L.V. King, "Electomagnetic Shielding at Radio Frequencies," Phil. Mag. J. Sci. Vol. 15, No 97, pp 201-223, Feb 1933. King's transfer function can be reduced to the form given in Equation 3.68.

of determining the time-domain response of the field inside a cylindrical or spherical enclosure is to find first the delta-function response h(t) of the structure. The response to any transient field g(t) instead of a delta-function is then obtained from:

$$r(t) = \int_{0}^{t} h\ (T)\ g\ (t - T)\ dT \tag{3.69}$$

When h(t) is known, it is generally somewhat simpler to evaluate, either analytically or numerically, the integral in Equation 3.69, than to obtain the Fourier transform of g(t), and then perform the inverse Fourier transform operation as discussed before. Thus, for example, the magnetic field r(t) inside a spherical or cylindrical enclosure, when the incident electric field is a delta function is given, from Equation 3.63, by:

$$r(t) = \frac{1}{2\pi} \int_{-\infty}^{\infty} \frac{-2e^{i\omega t} d\omega}{\frac{\mu_1}{\mu_2}\ \beta_2 b\ \text{Sin}\ \beta_2\ (a - b)\ -\ 2\ \text{Cos}\ \beta_2\ (a - b)} \tag{3.70}$$

If one denotes:

$$B = (a - b) \sqrt{\mu_2\ \sigma_2}$$

$$C = 2(\mu_2/\mu_1)\ (a - b)/b$$

where μ_2, σ_2 are the permeability and conductivity of enclosure material, one may write*:

$$r(t) = \frac{2C}{B^2}\ G(t/B^2) \tag{3.71}$$

where 2CG is plotted in Figure 3.14 as a function of (t/B^2) for various values of C. In Figure 3.14, if the abscissa is multiplied by B^2 and the ordinate is divided by B^2, one obtains a direct delta-function response of the magnetic field inside a cylindrical or spherical enclosure.

The EMP field inside the enclosure, as discussed above, will couple into electronic systems or components located within the enclosure. The EMP susceptibility of components or systems will depend on the effect of such a coupling.

---

* For evaluation of the integral, refer to "Electromagnetic Pulse Energy Coupling Studies," Vol. 1, American Nucleonics Corporation, ANC 30R-7, Feb 1967

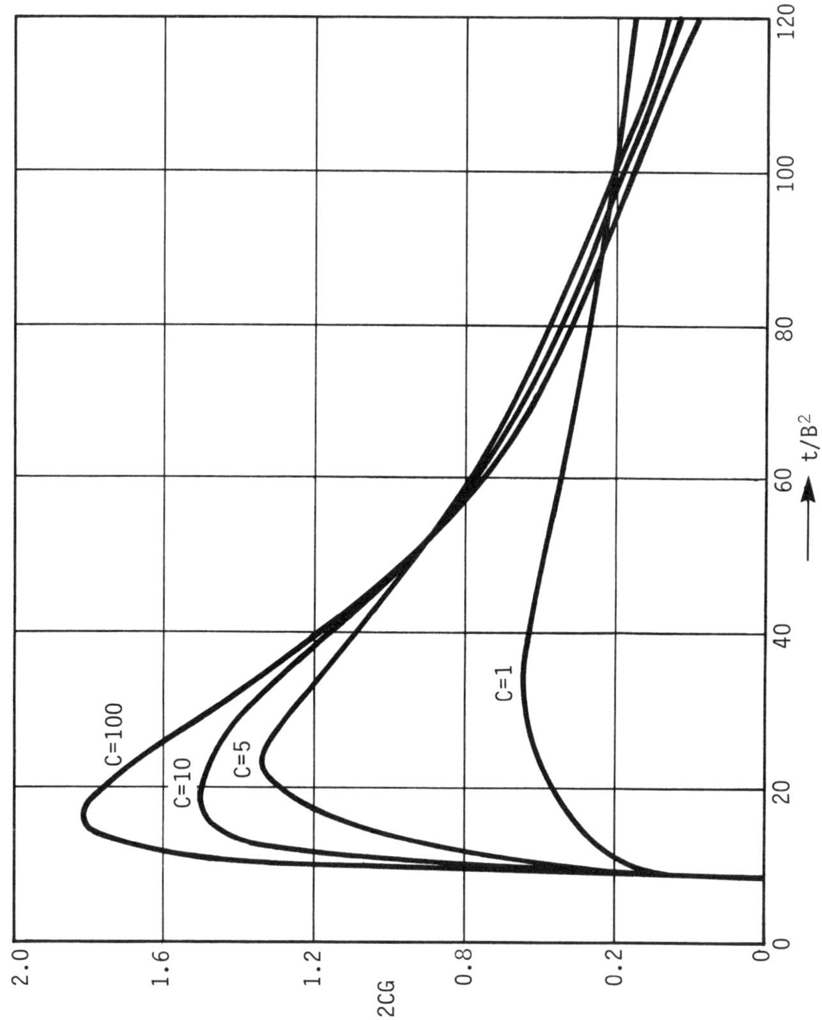

Figure 3.14 - Parametric Data for Delta Function Response of Cylinders and Spheres as Shielding Enclosures.

3.5 Coupling Through Slots and Slits in an Enclosure

The ideal electromagnetic shield which can prevent or substantially reduce EMP coupling to electronic equipment of concern is a completely closed, highly conducting shell surrounding the equipment. But invariably small apertures are present in shielding enclosures in most practical situations. A degradation of shielding effectiveness results from such apertures on the enclosure-wall. To obtain a measure of this degradation, one needs to solve a complex electromagnetic boundary value problem.

Conceptually, the mechanism of EMP coupling through an aperture on the enclosure wall may be understood by considering the aperture as a radiating source which produces an electromagnetic field inside the enclosure. The excitation of this aperture-radiating source occurs when an arbitrary EMP field impinges on the surface of the enclosure-wall containing the aperture. As one may expect, the field due to the aperture inside the enclosure will depend on the incident EMP field strength at the aperture, the dimensions of the aperture in terms of the wavelength of the impinging field, and the corresponding dimensions of the enclosure.

An exact solution of the field inside the enclosure due to the coupling through an aperture on the enclosure wall is very involved, particularly for arbitrary shapes and sizes of the aperture and enclosure. One can, however, characterize such a field by solving similar, but substantially simplified, problems, and by understanding the relative importance of various parameters that affect the coupling through the aperture.

The parameters and conditions that can affect the analytical solution of the field penetration through the aperture can assume a wide range, such that a number of different boundary value problems may have to be solved for understanding the overall implications of aperture coupling. Various factors that can influence the formulation and solution of the boundary value problems associated with the aperture coupling include:

- The polarization and orientation of the incident field exciting the aperture
- The finite dimensions of the enclosure wall containing the aperture
- Material characteristics of the enclosure walls
- Time-varying characteristics of the source such as EMP
- Aperture location and orientation with respect to enclosure axis and incident field
- Aperture dimension
- Geometry of the enclosure

It is evident, therefore, that all the aforementioned arbitrariness cannot be accommodated in a single formulation of the analytical problem on the aperture coupling.

In spite of the complexity of the situation that one confronts when attempting to determine the aperture coupling, useful information can be obtained in many cases by combining the results of specific, although simplified, solutions with engineering judgments. For example, if the enclosure dimension is so small that the propagation of natural modes for the geometry of the enclosure and frequencies of the source field of concern is inhibited, the field inside the enclosure will attenuate very rapidly with distance away from the aperture. Thus, if an aperture is located at the lateral surface of a cylindrical enclosure, and the enclosure wall extends at least 2 or 3 times the height of the aperture on either side, the field inside the enclosure will remain approximately the same whether the cylinder is finite or infinite in length, or whether it is open or closed at the ends. Furthermore, if the equivalent radius of the enclosure, in terms of the lowest wavelength of concern, is much smaller than unity, the time-varying nature of the field inside the enclosure will be very nearly the same, in most cases, as that of the incident field. The magnitude of the field inside the enclosure, however, will be reduced by a scale factor depending on the aperture dimensions.

An analysis of the physical problem on the aperture coupling also indicates that the material characteristics of the enclosure wall do not play a major role in the coupling mechanism as long as the walls of the enclosure are made of highly conducting materials. Thus, for the formulation of the problem, one can assume an infinitely conducting wall for the enclosure without introducing any serious error.

The effective electromagnetic source of nuclear origin is, in general, very complex in character. In most cases of interest, however, the source will be far away from the enclosure and one can assume a plane propagating wave due to EMP as it impinges on the enclosure wall.

3.5.1 Estimated Field Coupling Through an Aperture

In view of the above, it appears that a significant insight on the aperture coupling problem is achievable if one determines the coupled field inside an infinitely long, hollow, perfectly-conducting cylinder with an arbitrary radius, when an arbitrary electromagnetic field impinges on the cylinder wall and the aperture. The field inside the cylinder can always be expressed in terms of two independent mode sets corresponding to the axial E and H fields, regardless of the source of excitation, which is the aperture in this case. Therefore, the external field which impinges on the aperture then could also be resolved into two field components - one derivable from an E field which is parallel to the cylinder axis, and the other derivable from an H field also parallel to the cylinder axis.

Figure 3.15 shows an infinitely long cylindrical shell containing an aperture. The aperture is excited by a plane wave having arbitrary time-varying characteristics. The excited aperture, in turn, acts as a secondary source to illuminate the inside space of the shell.

If one assumes that the aperture is excited by an external H-field parallel to the axis of the cylindrical shell, the field inside the shell can be determined from a potential function Ψ^m which satisfies the wave equation:

$$\nabla^2 \Psi^m + \beta^2 \Psi^m = 0 \qquad (3.72)$$

β being the propagation factor, and is equal to $2\pi/(\text{wavelength})$. A time factor of $e^{i\omega t}$ is assumed in Equation 3.72.

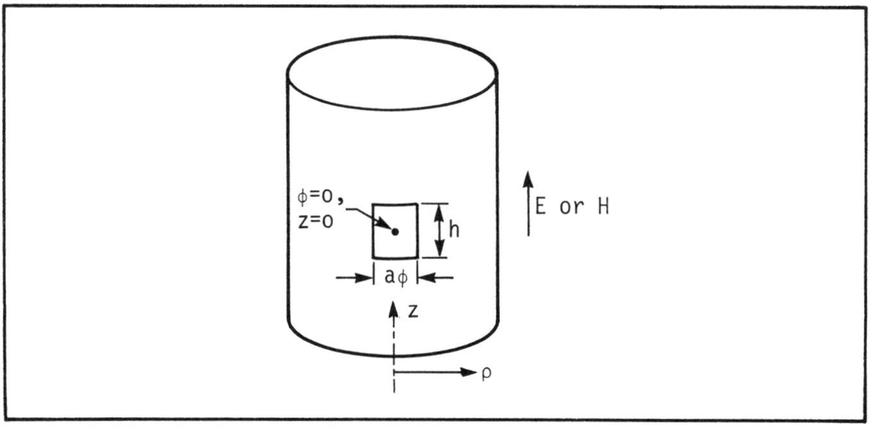

Figure 3.15 - An Aperture in a Hollow Cylinder.

The electric and magnetic field components corresponding to ψ^m are:

$$\vec{H} = \beta^2 \vec{\pi}^m + \nabla(\nabla \cdot \vec{\pi}^m)$$
$$\vec{E} = -i\omega\mu\,(\nabla \times \vec{\pi}^m) \qquad (3.73)$$

where:

$\vec{\pi}^m = \vec{u}_z \Psi^m$

$\vec{u}_z$ = unit vector along the axis of the cylinder

μ = magnetic permeability

and, $\omega = 2\pi \times$ frequency

For this particular case, where the axial field is only magnetic, $E_z = 0$ and the boundary conditions which the potential Ψ^m must satisfy are:

(1) E_ϕ is continuous at the aperture and is zero elsewhere on the cylinder shell surface

(2) H_z is continuous at the aperture (3.74)

If the incident field is $H_z{}^i$, one may also write this field in terms of a potential Ψ^i, such that $H_z{}^i = \beta^2 \Psi^i$, as is evident from Equation 3.73. This incident field and the cylindrical shell also create a scattering potential Ψ^s. In terms of the potentials, the boundary conditions given in Equation 3.74, become:

$$\Psi^i + \Psi^s = \Psi^m \quad \text{at the aperture}$$

$$\frac{\partial}{\partial \rho}(\Psi^i + \Psi^s) = \frac{\partial}{\partial \rho}\Psi^m \quad \text{at the aperture} \tag{3.75}$$

$$= \text{o elsewhere on the cylindrical surface.}$$

The potential Ψ^m, subject to the boundary condition that at the inner surface of the cylinder the tangential E-field should be zero, can be expressed* in terms of the aperture field as:

$$\Psi^m = \frac{1}{4\pi^2} \sum_{n=-\infty}^{\infty} e^{in\phi} \int_{-\infty}^{\infty} \frac{J_n(x\rho)}{x\, J_n'(xa)}\, G_n(\eta)\, e^{i\eta z}\, d\eta \tag{3.76}$$

where: $J_n(x\rho)$ = Bessel function of $(x\rho)$ of order n

$$J_n'(xa) = \frac{\partial}{\partial(x\rho)} J_n(x\rho)\Big|_{\rho = a}$$

$$x^2 = \beta^2 - \eta^2$$

and,

$$G_n(\eta) = \iint_{\text{Aperture}} \left[\frac{\partial \Psi^m}{\partial \rho}(\rho, \phi', z')\right]_{\rho = a} e^{-in\phi'}\, e^{-i\eta z'}\, d\phi' dz'$$

Although the field inside the cylindrical shell can be derived directly from Equations 3.73 and 3.76, Equation 3.76 is actually an integral equation rather than a solution of the potential function Ψ^m since Ψ^m also appears

---

* R. N. Ghose, "Diffraction Due to an Aperture on the Wall of a Shielded Enclosure," T.M. 30-5, American Nucleonics Corporation, July 1965

within the integral at the right hand side of Equation 3.76. The integral equation can be solved by utilizing the boundary condition at the aperture as shown in Equation 3.75. Ghose* has derived the following approximate expression for the magnetic field, H_z, at the axis of the cylindrical shell, when the width of the aperture ($a\Phi_o$) is small and when the external field impinging on the aperture is an H-field parallel to the axis of the cylinder:

$$H_z(0,0,z) \simeq H_z^i \frac{\Phi_o}{\pi} \sum_{s=1}^{\infty} \frac{1}{a J_o(\lambda_{os} a)} \frac{1}{\lambda_{os}^2 - \pi^2/h^2} \times$$

$$\left\{ \lambda_{os} e^{-\gamma_{os} h/2} \operatorname{Cosh} \gamma_{os} \bar{z} - \frac{\bar{z}}{h} \operatorname{Cos} \frac{\pi \bar{z}}{h} \right\} \qquad (3.77)$$

$$\text{for} \quad |\bar{z}| \leq h/2$$

where:

$$J_o'(\lambda_{os} a) = 0, \quad \text{for } s = 1, 2, 3 \ldots$$

$$\bar{z} = z - z_o$$

z_o being the z-coordinate at the center of the aperture,

$$a\Phi_o = \text{width of the aperture}$$

and,

$$\gamma_{os}^2 = \lambda_{os}^2 - (2\pi/\lambda)^2$$

λ being the wavelength of the incident field.

For $|\bar{z}| > h/2$, the second bracket in Equation 3.77 is replaced by:

$$\lambda_{os} e^{-\gamma_{os}\bar{z}} \operatorname{Cosh} \left[\frac{\gamma_{os} h}{2} \right] \qquad (3.78)$$

The corresponding radial component of the magnetic field, also generated by the same incident H-field (parallel to the axis of the cylinder) is given by:

---

* R. N. Ghose, "Electromagnetic Pulse Energy Coupling Studies," Vol. 1, American Nucleonics Corporation, ANC 30R-7, Glendale, California, February 1967

$$H_\rho(o, o, z) = H_z^i \frac{\Phi_o}{\pi} \sum_{s=1}^{\infty} \frac{1}{1 - \frac{1}{\lambda_{\ell s}^2 a^2}} \frac{\lambda_{\ell s}}{J_1(\lambda_{\ell s} a)} \times \tag{3.79}$$

$$\left\{ \frac{e^{-\gamma_{\ell s} h/2} \operatorname{Sinh}(\gamma_{\ell s} \bar{z}) - \operatorname{Sin} \frac{\pi \bar{z}}{h}}{\lambda_{\ell s}^2 a^2 + \pi^2 a^2/h^2} \right\}$$

where: $\bar{z} = z - z_o$, and $|z - z_o| \leq h/2$

When $|\bar{z}| > h/2$, the second bracket in Equation 3.79 is replaced by

$$e^{-\gamma_{\ell s} \bar{z}} \operatorname{Sinh}(\gamma_{\ell s} h/2) \tag{3.80}$$

Also in Equation 3.78, $J_1'(\lambda_{\ell s} a) = 0, s = 1, 2, 3 \ldots,$

and, $\gamma_{\ell s}^2 = \lambda_{\ell s}^2 - (2\pi/\lambda)^2$

λ being the wavelength of the incident field.

Figure 3.16 shows H_z (0, 0, z) component of EMP coupling through an aperture on the wall of a cylindrical shell for various values of h and Φ_o.

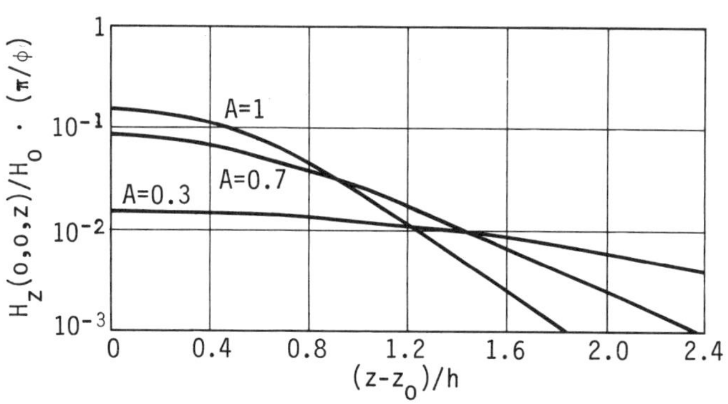

Figure 3.16 - Variation of Axial Magnetic Field Coupled Inside a Cylinder by an Aperture.

Similarly, Figure 3.17 shows the corresponding H_ρ (0, 0, z) component. In both cases the incident field is assumed to be a magnetic field parallel to the cylinder axis.

An examination of Figure 3.16 indicates that the coupled magnetic field at the axis of the cylinder approaches almost zero at z_o corresponding to the center of the aperture. Also the field at the shadow zone, corresponding to $|\bar{z}| > h/2$, attenuates rapidly with increasing value of z, the maximum coupled field being for the values of z corresponding to the upper and lower edges of the aperture.

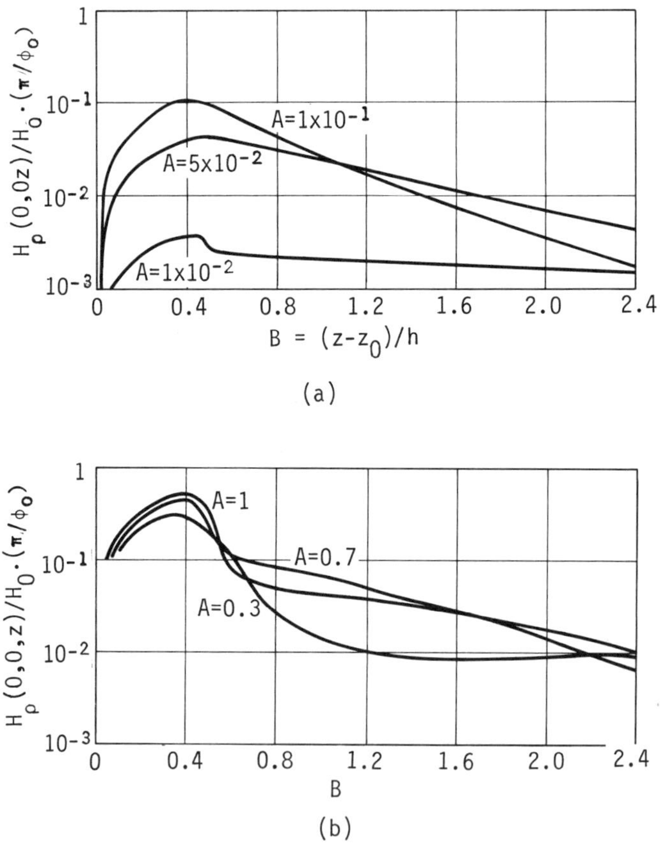

Figure 3.17 - Variation of Radial Magnetic Field Coupled by Aperture. (a)For A=(h/a)=0.1,0.05 and 0.01. (b) For a=(h/a)=1, 0.7 and 0.3.

3.5.2 Correlation of Theoretical and Experimental Results

Bombardt* obtained experimental results on the coupling of magnetic field at the axis of a circular cylindrical shell due to an aperture, at the cylinder surface. Here the aperture, which was in the form of a circular hole as shown in Figure 3.17, was excited by an axial magnetic field impinging on the cylinder surface. Figure 3.18 shows H_z (0, 0, z) component of the measured coupled field. It is seen in this case that the axial magnetic field approachs zero at the value of z corresponding to the center of the aperture; then it reaches a peak value for z corresponding to the edges of the aperture, and finally attenuates rapidly in the shadow zone. Although the impressed field in Bombardt's experiment was in the form of that due to EMP, that is $H_z^i(t) = \beta t e^{-\alpha t}$, α and β being constants, the nature of the field variation along the axis, as shown in Figure 3.18, would be very nearly the same for the sinusoidal-impressed field. Also, for a small aperture, the nature of the field variation along the axis is very similar to that obtained by the quasi-static approach formulated by Kaden**.

The aperture-coupled field inside a cylindrical shell when the impinging field at the aperture is an E-field parallel to the axis of the cylinder has been

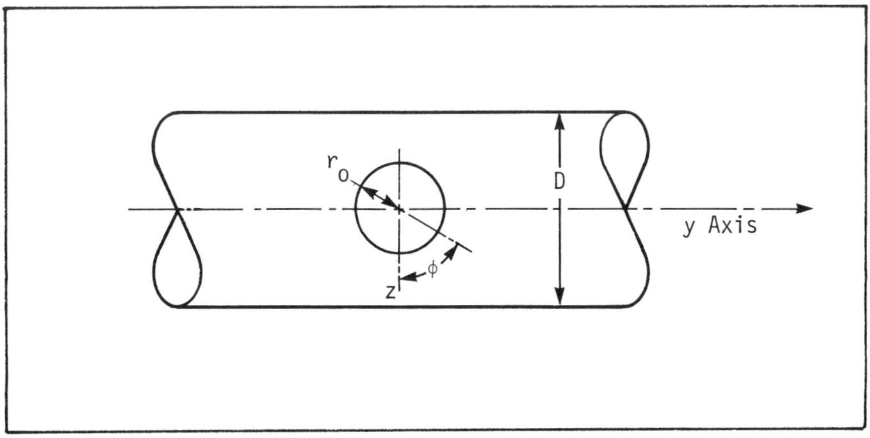

Figure 3.18 - A Circular Aperture On A Hollow Cylinder

* J. N. Bombardt, "Quasi-Static Magnetic Field Transmission Through Circular Apertures", Note 9, EMP Interaction Notes, Air Force Weapons Laboratory, Sept 1966

** Heinrich Kaden, "Wirbelstorme und Schirmung In Der Nachaicentechnik," Technische Physik in Einzeldarstellungen 10, Springer-Verlag 1959. Theoretical results following Kaden's approach are shown in Figure 3.19 for comparison.

analyzed by Ghose*. From the results of theoretical analyses and experimental data on the aperture-coupled field, it is seen that the smaller the aperture, the less the magnitude of the field inside the enclosure. Also when $2\pi a/\lambda$ is much less than 1, a being the radius of the cylinder, the field derived from the quasi-static approach (equivalent to $\beta = o$) is usually a good estimate of the field to be encountered actually.

When the incident field is a magnetic field parallel to the axis of the cylinder, the ρ-component of the aperture-coupled magnetic field along the axis is considerably higher than the corresponding z-component even though H_z is the dominant field outside the cylinder. For small values of βa, the transient response of the field at any point inside the enclosure is very nearly the same as that of the incident field, excepting that the magnitude of the coupled field is reduced in proportion to the aperture dimension.

The expected values of the aperture-coupled field inside an enclosure, due to most apertures to be encountered in real life, seem to be lower by several orders of magnitude than the incident field.

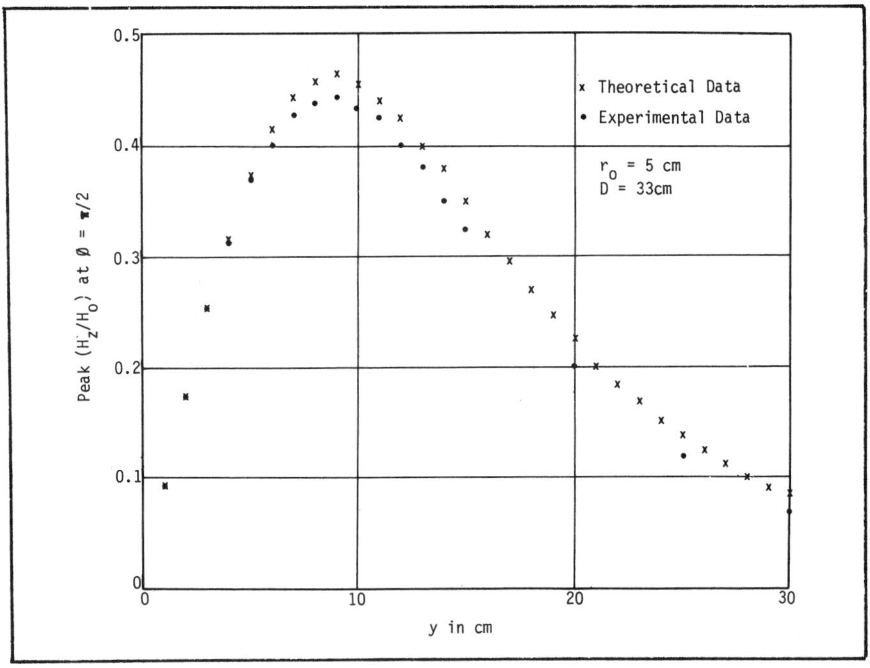

Figure 3.19 - Axial Magnetic Field Coupled by Aperture.

* opt. cit.

3.6 Coupling Through Cables and Transmission Lines

The cables which interconnect parts of an electronic or weapon system can couple damaging amounts of energy into the system if they are exposed to the EMP environment. The amount of EMP energy thus coupled, particularly its effect on the cable load in the form of load current and voltage and their time history, depends on the cable configuration, characteristics of the exposed cable, and the mechanism by which the coupling is effected. Important factors to be considered in this regard are the geometry of the cable run, orientation and magnetic field vectors of EMP and terminations of the conductors.

From the viewpoint of EMP coupling, two broad classes of cable problems need to be considered. They are:

- Long-exposed or underground cable, where the EMP environment varies over the length of the cable
- Short-exposed cable, where the EMP environment is constant over the length of the cable.

3.6.1 Long Cables

Common long cables of concern are bare conductors or insulated cables at or near the surface of the earth and insulated subsurface cables. The conventional uniform transmission-line theory can be used to determine the EMP coupling.

Let a bare conductor cable running parallel to the earth's surface be considered first. Here the ground serves as the return conductor. Such a line has a series impedance per unit length Z, and a shunt admittance per unit length Y. The distributed generator voltage impressed per unit length of the cable will be due to the radial electric field of EMP along the cable. At each point along the conductor, a current increment is injected through the admittance Y by this radial electric field component. The current combines with currents arriving from other parts of the cable, with proper regard for phase, and propagates along the line.

The EMP-induced current at any point x along the line satisfies the differential equation:

$$\frac{d^2 \tilde{I}(x, \omega)}{dx^2} + \tilde{\Gamma}^2 \tilde{I}(x, \omega) = Y(\omega)\, \tilde{E}(x, \omega) \qquad (3.81)$$

where $\tilde{I}(x, \omega)$ is the Fourier transform of the current, $\tilde{I}(x, t)$, $\tilde{\Gamma}^2 = Z/Y$ and $\tilde{E}(x, \omega)$ is the Fourier transform of the radial electric field component along the conductor due to EMP. Here x, ω and t denote respectively a point along the cable under consideration, angular frequency (or transform parameter) of the EMP field, and time. If the wire is infinitely long, the solution of (3.81) is:

$$\tilde{I}(x, \omega) = K_1 e^{\tilde{\Gamma} x} + K_2 e^{-\tilde{\Gamma} x} + \frac{Y}{2\tilde{\Gamma}} \int_{-\infty}^{\infty} \tilde{E}(x', \omega) e^{-\tilde{\Gamma}|x-x'|} dx' \tag{3.82}$$

If the reflection terms are neglected, Equation 3.82 reduces to:

$$\tilde{I}(x, \omega) = \frac{\tilde{E}(x, \omega)}{Z} \tag{3.83}$$

where Z is the series impedance per unit length of the conductor and ground, and $\tilde{I}(x, \omega)$ is the current induced in the bare conductor at x. To compute $\tilde{I}(x, \omega)$, one may write $Z = Z_s + Z_w$, where Z_s and Z_w are respectively the surface impedance and the impedance of the wire. Expressions for Z_s and Z_w, as obtained from the solution of the boundary-value problem are:

$$Z_s = \frac{\beta_1}{2\pi r_c \sigma_g} \frac{H_o^{(1)}(\beta_1 r_c)}{H_1^{(1)}(\beta_1 r_c)}$$

$$Z_w = \frac{\beta_c}{2\pi r_c \sigma_c} \frac{J_o(\beta_c r_c)}{J_1(\beta_c r_c)} \tag{3.84}$$

where:

$\beta_1 = \left[\omega^2 \mu \varepsilon_g - i\omega\mu\sigma_g\right]^{1/2}$ = propagation constant in the earth

ε_g = dielectric constant of the earth

σ_g = conductivity of the earth

$\beta_c = \left[\omega^2 \mu \varepsilon_o - i\omega\mu\sigma_c\right]^{1/2}$ = propagation constant in the conductor

σ_c = the conductor conductivity

μ = permeability of free space

r_c = radius of the wire

$H_o^{(1)}(x)$ = Hankel function of the first kind and order zero of x

$H^{(1)}(x)$ = Hankel function of the first kind and order one of x

$J_o(x)$ = Bessel function of order zero of x

$J_1(x)$ = Bessel function of order one of x

Assuming a high conductivity of the metal and ground, one can use the asymptotic expression for Bessel and Hankel functions and write:

$$Z \simeq \frac{x}{2\pi r_c \sigma_c} \left[\frac{J_o(x)}{J_1(x)} + \frac{x}{2} \ln \left| \frac{x^2 a}{4} \right| \right] \tag{3.85}$$

where,

$$x = \left[i\omega\mu\sigma_c r_c^{\,2} \right]$$

$$a = \frac{\gamma^2 \sigma_g}{\sigma_c}, \quad \gamma = 1.781 \text{ --}$$

For a long insulated conductor at or near the surface of the earth, the electric field coupling is predominantly through the insulation capacitance. Characteristics of propagation of the EMP-induced current are determined by the insulation, and the induced current in this case can propagate over long distances because of the relatively low line-loss.

3.6.2 Induced Current in a Subsurface Insulated Cable

EMP energy can be carried to a system by underground cables, the degree of energy coupling in such cases being dependent on the characteristics of the underground cable and the EMP field inducing the current in the cable. Since the transmission-line theory is valid for underground cables subjected to the EMP field, one may begin to derive an expression for the induced current as the solution of Equation 3.81. Thus,

$$I(x, \omega) = K_1 e^{\Gamma x} + K_2 e^{-\Gamma x} + F(x, \omega) \tag{3.86}$$

where,

$$F(x, \omega) = -\frac{1}{2Z_o} \left[e^{\Gamma x} \int e^{-\Gamma x'} E(\omega, x')\, dx' - e^{-\Gamma x} \int e^{\Gamma x'} E(\omega, x')\, dx' \right]$$

$$Z_o = (Z/Y)^{1/2}$$

and K_1 and K_2 are arbitrary constants. The corresponding line voltage can be written as:

$$V = -Z_o \left[K_1 e^{\Gamma x} - K_2 e^{-\Gamma x} + F_v(x, \omega) \right] \qquad (3.87)$$

where: $$F_v(x, \omega) = \frac{1}{\Gamma} \left[\frac{\partial F(x, \omega)}{\partial x} \right]$$

The constants K_1 and K_2 can be evaluated from the boundary conditions. Thus, for a line extending from x = a to x = b and terminating at $Z_o Z_a$ at x = a and $Z_o Z_b$ at x = b, one may write:

$$V(a) = -I(a) Z_a Z_o$$

$$V(b) = I(b) Z_b Z_o$$

and,

$$K_1 = \frac{e^{-\Gamma b}(1 - Z_b)\left[F_v(a) - Z_a F(a)\right] - e^{-\Gamma a}(1 + Z_a)\left[F_v(b) + Z_b F(b)\right]}{e^{\Gamma(b-a)}(1 + Z_a)(1 + Z_b) - e^{-\Gamma(b-a)}(1 - Z_a)(1 - Z_b)}$$

(3.88)

$$K_2 = \frac{e^{\Gamma b}(1 + Z_b)\left[F_v(a) - Z_a F(a)\right] - e^{\Gamma a}(1 - Z_a)\left[F_v(b) + Z_b F(b)\right]}{e^{\Gamma(b-a)}(1 + Z_a)(1 + Z_b) - e^{-\Gamma(b-a)}(1 - Z_a)(1 - Z_b)}$$

To obtain an expression for the current, one has yet to evaluate Γ. For the underground cable with insulation,

$$\Gamma^2 = -i\omega C (R - i\omega L) \qquad (3.89)$$

where: $$C = \frac{2\pi K_1 \varepsilon_o}{\ln\left[\frac{r_c + T}{r_c}\right]}$$

$$R = \frac{1}{\pi r_c^2 \sigma_c}$$

$$L = L_{\substack{bare\\wire}} + \frac{\mu_o}{2\pi} \ln\left[\frac{r_c + T}{r_c}\right]$$

r_c = wire radius

$_c$ = wire conductivity

T = insulation thickness

$L_{\substack{bare\\wire}}$ = inductance per unit length of bare wire or radius r_c in the earth

μ_o = $4\pi \times 10^{-7}$ henry/meter

K_1 = dielectric constant of the insulation

ε_o = 8.854×10^{-12} farad/meter

From Equations 3.86, 3.88 and 3.89, one can determine I (x, ω) for the EMP excitation in the form of F(a), F(b), F_v(a) and F_v(b). It may be noted that for a matched cable, where $Z_a = Z_b = 1$,

$$K_1 = \frac{-e^{-\Gamma b}}{2}\left[F_v(b) + F(b)\right]$$

$$K_2 = \frac{e^{\Gamma a}}{2}\left[F_v(a) - F(a)\right] \tag{3.90}$$

The time-domain response of the EMP-induced current is obtained from the Fourier transform of I(x, ω) in the time domain.

3.6.3 Short Cables

For long cables considered so far, the EMP environment may be different along the length of the line. EMP-induced driving functions F(a), F(b), F_v(a) and F_v(b) account for the variation of the EMP environment along the line, since they involve integration of $E(\omega, x')e^{-\Gamma x'}$ and $E(\omega, x')e^{\Gamma x'}$ over the entire length of the line as shown in Equations 3.86 and 3.87. In contrast, the EMP environment is assumed uniform or constant over the length of short cables.

Let a short coaxial cable immersed in an incident electric field E_i parallel to its axis be considered first. With the outer conductor considered as a linear antenna, the external incident field E_i induces an axial current distribution on this conductor. If the current $I_s(x)$ is known as a function of x, the coupling into the interior of the coaxial cable can be expressed as a continuous distribution of incremental generators with voltage $Z_T I_s(x)\,dx$, for each element dx, where Z_T is the transfer impedance of the outer shield.

Let it be assumed that the length of the cable under consideration is h, and the cable is terminated in impedances Z_1 and Z_2 as shown in Figure 3.20. If $V_1(x, \xi)$ and $I_1(x, \xi)$ denote the voltage and current at point x when a point source V_g is impressed at $x = \xi$ in series with the line, Schelkunoff* has shown:

$$\begin{aligned} D\,I_1(x, \xi) &= V_g f(x)\left[Z_o \operatorname{Cos} \beta(h-\xi) + i\,Z_2 \operatorname{Sin} \beta(h-\xi)\right], \quad x < \xi \\ &= V_g g(h-x)\left[Z_o \operatorname{Cos} \beta\xi + i\,Z_1 \operatorname{Sin} \beta\xi\right], \quad x > \xi \end{aligned} \tag{3.91}$$

where:

$$f(x) = Z_o \operatorname{Cos} \beta x + i\,Z_1 \operatorname{Sin} \beta x$$

$$g(h-x) = Z_o \operatorname{Cos} \beta(h-x) + i\,Z_2 \operatorname{Sin} \beta(h-x)$$

$$D = Z_o\left[Z_o(Z_1 + Z_2) \operatorname{Cos} \beta h + i\,(Z_o^2 + Z_1 Z_2) \operatorname{Sin} \beta h\right]$$

β = propagation phase constant of the line

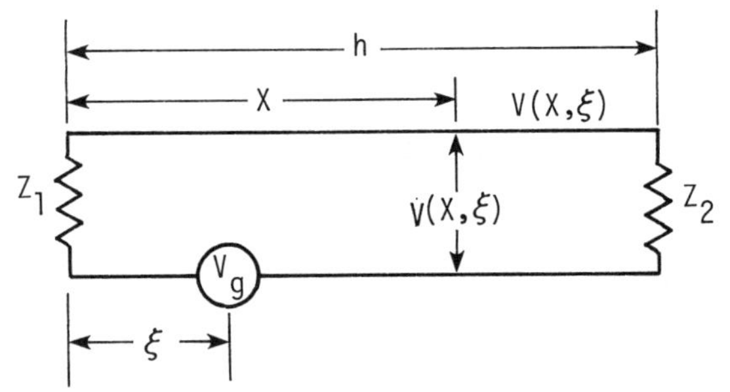

Figure 3.20 - A Section Of Cable Energized By a Series Generator At A Point Along The Line.

* S. A. Schelkunoff, "Electromagnetic Waves," D. Van Nostrand Co., Inc., New York, 1943, p. 201.

$$V_g = Z_T I_s(\xi)\, d\xi$$

$I_s(\xi)$ = sheath current as a function of ξ

The current I(x) can now be evaluated by superposition of the contributions for various ξ, that is:

$$I(x) = \frac{Z_T}{D} g(h-x) \int_o^x \left[Z_o \operatorname{Cos} \beta\xi + i Z_1 \operatorname{Sin} \beta\xi\right] I_s(\xi)\, d\xi$$

$$+ \frac{Z_T}{D} f(x) \int_x^h \left[Z_o \operatorname{Cos} \beta(h-\xi) + i Z_2 \operatorname{Sin} \beta(h-\xi)\right] d\xi \tag{3.92}$$

Evaluating I(x) at x = 0, the current at the load $Z_1 = Z_L$ is obtained as:

$$I(o) = I_L = \frac{Z_o Z_T}{D} \int_o^h \left[Z_o \operatorname{Cos} \beta(h-\xi) + i Z_2 \operatorname{Sin} \beta(h-\xi)\right] I_s(\xi)\, d\xi \tag{3.93}$$

For numerical evaluation of I_L, one must know Z_T and the distribution of $I(\xi)$. Z_T can be determined analytically or experimentally. For the distribution of $I(\xi)$, one may treat the outer conductor as an unloaded receiving antenna in a uniform field. The induced-sheath current has the following leading term:

$$I_s(x) = I_s(o) \left[\frac{\operatorname{Cos} \beta_o x - \operatorname{Cos} \beta_o h}{1 - \operatorname{Cos} \beta_o h}\right] \tag{3.94}$$

where: β_o = propagation constant of the medium surrounding the antenna

$I_s(o)$ = $-h_e E_i(\omega)/Z_a$

h_e = effective length of a monopole antenna of actual height h

Z_a = input impedance of the antenna

Substituting the expression of $I_s(x)$ as given in Equation 3.94 into Equation 3.93, and carrying out the integration, one obtains:

$$I_L(\omega) = \frac{I_s(o)\, Z_T Z_o}{D\,(1-\cos\beta_o h)} \left[\frac{i \cos\beta_o h (Z_2 \cos\beta h - Z_2 + i Z_o \sin\beta h)}{\beta} + Z_o \left(\frac{-\beta \sin\beta h + \beta_o \sin\beta_o h}{\beta_o^2 - \beta^2} \right) + \frac{i\beta Z_2 (\cos\beta h - \cos\beta_o h)}{\beta_o^2 - \beta^2} \right] \tag{3.95}$$

Equation 3.95 is the expression for the current in the load $Z_1 = Z_L$ due to the incident field $E_i(\omega)$. To obtain the time-domain response, the inverse-Fourier transform operation is needed. Thus,

$$i_L(t) = \frac{1}{2\pi} \int_{-\infty}^{\infty} I_L(\omega)\, e^{i\omega t}\, d\omega$$

The integration is usually carried out by numerical integration techniques using a digital computer.

It is seen from Equation 3.95 that an important parameter determining $I_L(\omega)$ is the transfer impedance Z_T. An approximate expression for Z_T of a solid outer conductor is given by:

$$Z_T = \frac{\eta}{2\pi \sqrt{r_i r_o} \sinh \gamma T} \tag{3.96}$$

where, η = intrinsic impedance of the shield

γ = propagation constant in the shield

For a good conductor:

$$|Z_T| = R_o F(u)$$

where: R_o = dc resistance per unit length of the outer shield

$u = 2T/\delta$

δ = skin depth = $500/\sqrt{\sigma f}$ meter

σ = conductivity

f = frequency corresponding to ω

$F(u) = u\left[\text{Cosh } u - \text{Cos } u\right]^{-1/2}$

The phase of Z_T is given by:

$$\angle Z_T = \phi(u) = \frac{\pi}{4} - \tan^{-1}\left[\tan\frac{u}{2} / \tanh\frac{u}{2}\right]$$

Figure 3.21 shows a plot of F(u). Experimentally obtained values of Z_T for commonly used cables RG 62 B/U, RG 12/U Armored, and RG 13 A/U are shown in Figures 3.22, 3.23 and 3.24.

The foregoing discussions on the cable currents induced by an incident EMP field attempt to highlight the analytical problems involved in estimating currents in a few types of cables. They show parametric influences on the induced current due to cable characteristics. Types of cables and their configurations used in electronics or weapon systems are too numerous to deal with in this section. The highlights on the subject as discussed and the references cited at the end of this section, however, may be useful to estimate EMP coupling mechanism for other forms or types of cables of interest.

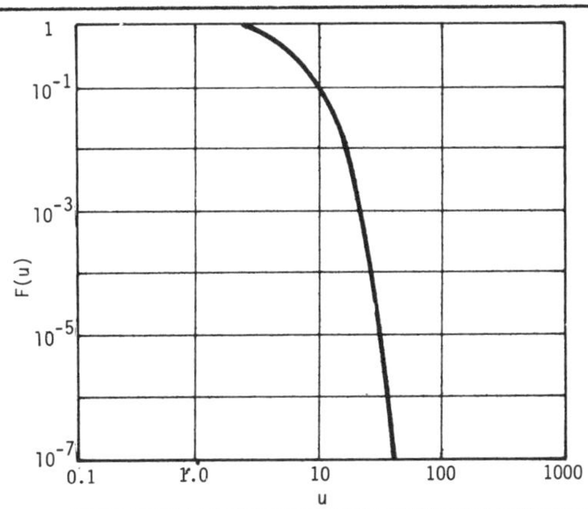

Figure 3.21 - A Plot of F (u) As a Function of u · F≈1 For u<1, $F \approx \left[2ue^{-u/2}\right]^{1/2}$ For u>5.

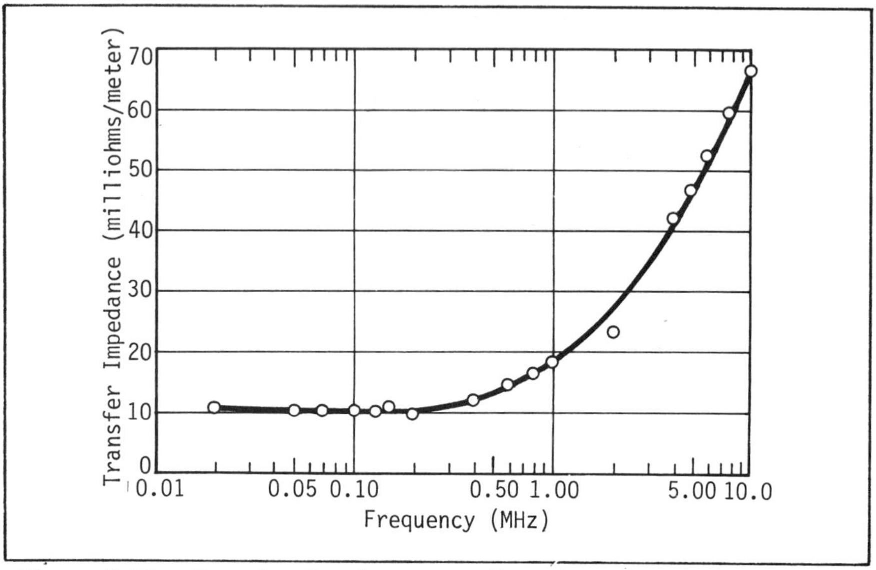

Figure 3.22 - Transfer Impedance as a Function of Frequency For RG 62 B/U.

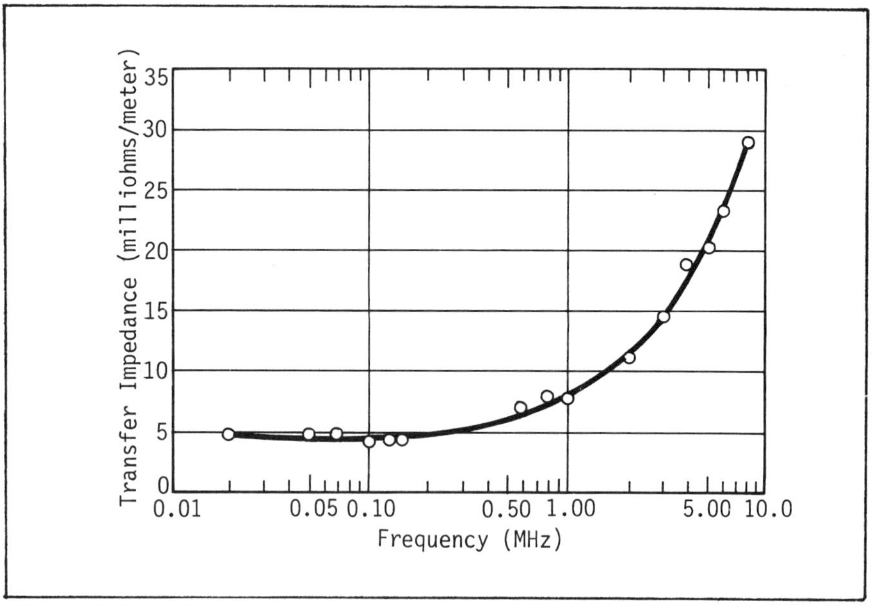

Figure 3.23 - Transfer Impedance as a Function of Frequency For RG 12/U Armored Cable (Armor Floating).

3.6.4 EMP Coupling Through Utility Systems

Utility systems such as telephone lines, power cables, gas and water pipes, etc., can also be the carriers of EMP-induced energy into buildings housing electronic equipment. A fraction of this energy may eventually couple to the sensitive equipment. The degree and nature of EMP couplings through telephone lines and power cables can be estimated from the discussions presented earlier in this section, since they are mostly in the form of long cables. Since these cables may be overhead wires or subsurface cables, EMP-induced currents will be different in each case because of different impedances involved. Appropriate terminations also have to be taken into account to determine the EMP-induced currents.

EMP couplings by gas and water pipes are usually not significant since such pipes are mostly underground bare conductors, and the induced current attenuates rapidly as it propagates along the line.

The actual EMP couplings from utility lines to sensitive electronic equipment of concern can be by conductive or radiative means. Conventional EMI concepts and techniques are applicable to estimate such couplings even when the nature of the waveform of the induced currents and their durations are far from conventional. For example, one can select a few representative frequencies from the spectrum of induced current $I(\omega)$ and calculate the

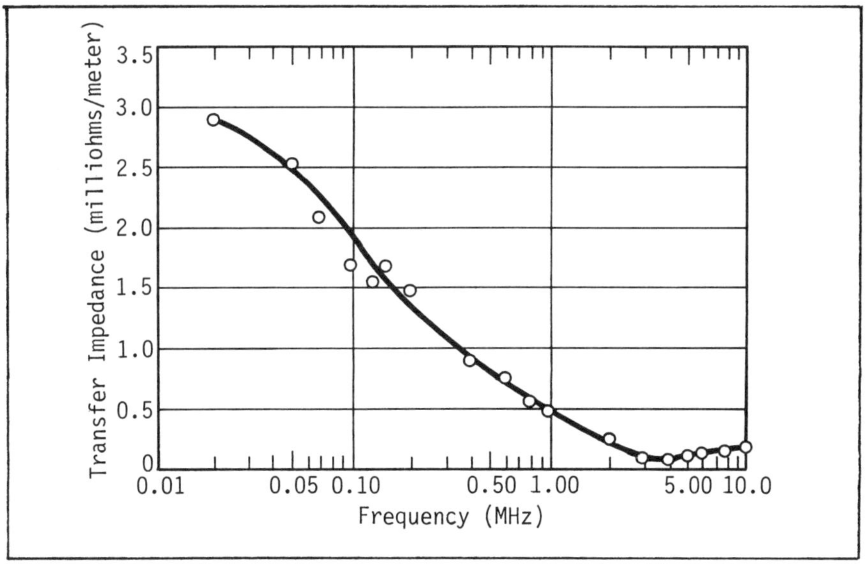

Figure 3.24 - Transfer Impedance as a Function of Frequency For RG13 A/U.

worst-case conductive and radiated fields at the point of concern by assuming a separate source for each frequency, the source strength being the amplitude of the induced current at the frequency.

It may be noted that exact or nearly exact estimates of EMP-induced currents or voltages are neither practical nor necessary, the primary object in most cases being to provide adequate protection to systems which could be vulnerable to EMP.

3.7 Hybrid Coupling Mechanism

A few representative means of EMP couplings to electronics or weapon systems have been considered in the previous sections. In many cases of practical interest, a coupling is effected by more than one means. Also, there could be couplings at multiple points by the same means, such as the EMP penetration inside an enclosure by multiple windows. The effect of multiple couplings can be taken into account by superposing the EMP-induced fields or voltages or currents due to different sources. This assumes that there are no mutual interactions among different means of couplings.

3.8 References

1. Schelkunoff, S. A., *Electromagnetic Waves*, D. Van Nostrand Co., Inc., New York, N.Y., 1943.

2. Brown, G. L., Ghose, R. N., Green, J. X., Putzer, E. J., Schwartz, K., Wouters, L. F., *Electromagnetic Pulse Energy Coupling Studies*, Volume 1, American Nucleonics Corporation, Glendale, Ca.,February 1967.

3. Kaden, Heinrich, *Wirbelstrome und Schirmung in der Nachrichtentechnik*, Springer-Verlag, 1959.

4. Ghose, R. N., *Diffraction Due to An Aperture on Wall of A Shielded Enclosure*, TM 30-5, American Nucleonics Corporation, Glendale, Ca., July 1965.

5. Bombardt, J. N., *Quasi Magnetic Field Transmission Through Circular Apertures*, Note 9, EMP Interaction Notes, Air Force Weapons Laboratory, September 1966.

6. Stogryn, A., *Theoretical Studies Relating to Space General Corporation Pulse Model, Part III, Induced Currents in Bare Wires*, Space General Corporation, SGC-726-TM-6, July 1966.

7. Stogryn, A., *Theoretical Studies Relating to Space General Corporation Pulse Model, Part IV, Induced Currents in Insulated Wires*, Space General Corporation, SGC-726-TM-6, January 1967.

8. Vance, E. F., *Internal Voltages and Currents in Complex Cables*, EMP Interaction Note VIII, Stanford Research Institute, June 1967.

9. Sunde, E. D., *Earth Conduction Effects in Transmission Systems*, D. Van Nostrand Co., Inc., New York, N.Y.1949.

10. Vance, E. F., *Electromagnetic Pulse Handbook for Electric Power Systems*, Stanford Research Institute, Proj. 2709 DNA 3466F, February 1975.

11. Sturm, R., *EMP-Induced Currents and Fields in Shelters*, Wehrwissenschaftliche Dienststelle der Bundeswehr fur ABC-Schutz, Third Symp. and Tech. Exhibition on Electromagnetic Compatibility, Rotterdam, May 1979.

12. Gallon, I. L., *EMP Coupling to Long Cables*, Atomic Weapons Res. Estab., England, Third Symp. and Tech. Exhibition on Electromagnetic Compatibility, Rotterdam, May 1979.

13. Aguet, M., Blech, Ph., Ianovici, M., and Lin, C. C., *Response Calculation on Transmission Cables*, Proc. I.E.E. Conference on EMC, Southampton, September 1980.

14. Grønhaug, K. L., *Calculations of Current Induced in Long Conductors by Exo-atmospheric EMP*, Fourth Symp. and Tech. Exhibition on Electromagnetic Compatibility, Zurich, March, 1981.

CHAPTER 4

EMP Susceptibilities in Electronic Systems

EMP-coupled energy may affect an electronic system performance temporarily or permanently. Temporary effects are those which disappear along with or shortly after the disappearance of EMP. For example, transient circuit disturbances resulting from EMP may be interpreted by the circuit as signals. Information errors may be introduced when such disturbances appear in the form of noise, deteriorating the signal-to-noise ratio for the circuit. Permanent effects in turn have lingering consequences. Circuit latch-up or burn-out of components are examples of permanent effects. Malfunctions due to both temporary and permanent effects could be of concern, and they are referred to as EMP susceptibilities. Typical susceptibilities in electronic systems due to EMP, and techniques and methods for susceptibility assessment and prediction, are outlined in this chapter.

4.1 Transient Circuit Upset

EMP transients which are picked up by an electronic system and propagated into the electronics or electrical circuits are generally short-duration, damped sine-waves with dominant frequencies, usually in the range of 100 KHz to 100 MHz. Most EMP-induced transients in electrical circuits are regarded as sources of temporary effects, although transients can lead to "latch-up" of a circuit or a semiconductor, thus requiring a reset before normal operating function can be restored. For example, flip-flops may be triggered by transients, counters can change count, memory can be altered due to driving current or direct magnetic field effect resulting from the transients, etc. Transients, as signals, may also be amplified in electronic circuits and the amplified signals may be interpreted as control signals by the circuits. Such control signals in turn may result in a system malfunction by providing controls where they are not needed or by altering the degree of control.

The sensitivity of digital circuits to upset by EMP transients depend on many factors such as:

- Magnitude of the "true" or "false" voltage level.
- Switching time of the active semiconductors (rise time, storage time, etc.).
- Presence of background noise in the circuit due to nearby noise sources.
- Time of arrival of EMP transient with respect to clocking waveform.
- Magnitude of resistance levels in circuits.
- Circuit inductances and capacitances.
- Magnitude of bias voltage in circuits.
- Schematic of the circuit and its physical layout.

Because of these numerous variables, it is difficult to present a comprehensive analysis or a methodology for the prediction of susceptibilities due to EMP transients. Some generalization, however, could be attempted regarding the response of digital circuits due to EMP transients. First, the most sensitive lead on a digital circuit is usually the input lead, although the ground lead or dc-power lead could be as sensitive or more sensitive in some cases. Secondly, one can anticipate that at low frequencies, the threshold of the circuit can be related to its normal operating levels, or logic voltage levels. If, for example, the circuit has a +6 volt "true" and a -6 volt "false" nominal operating voltage, the circuit may be expected to switch at 3 or 4 volts approximately, since circuits are designed to accommodate a reasonable variation of operating voltages without malfunctioning. Rarely, this enabling voltage will be much smaller than 3 to 4 volts or much larger than -3 to -4 volts. Figure 4.1 shows the threshold in terms of logic levels for this case. The regions of uncertainties for this circuit are +4 to +3 volts and -3 to -4 volts.

4.1.1 Electronic Circuit Susceptibility

The susceptibility of a circuit varies, in general, with frequency. If the input resistance of a transistor is shunted by a capacitance as shown in Figure 4.2, the sensitivity is expected to increase and the threshold is expected to increase at high frequencies. Most circuit thresholds seem to turn upward at or around 5 MHz. A series inductance will have the same effect as a shunt capacitance at high frequencies. A combination of a shunt capacitance and a series inductance will cause a steeper rise on the upset threshold curve than what is shown in Figure 4.2.

Another type of high-frequency effect is due to switching time. If, for example, the circuit cannot switch (turn on or turn off) in the time of a half a cycle of the EMP waveform, the threshold for upset will approach infinity asymptotically as shown in Figure 4.3.

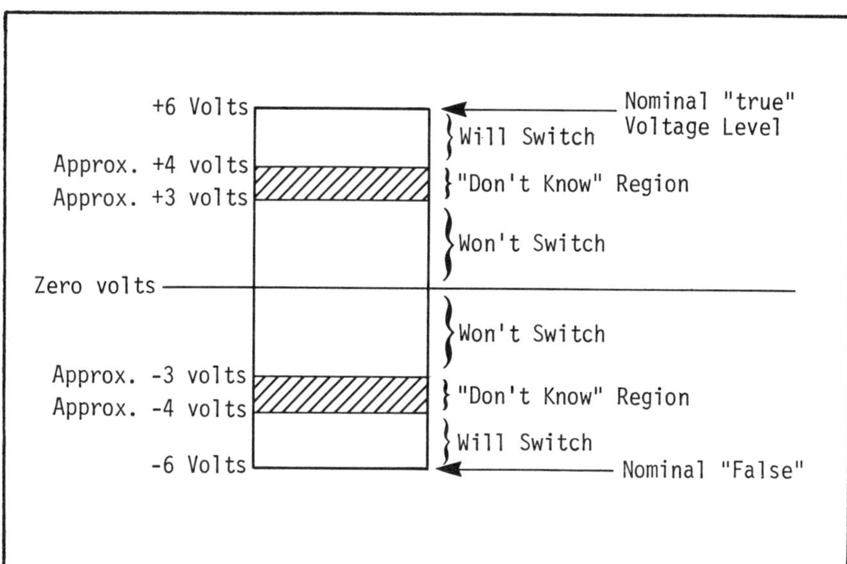

Figure 4.1 - Digital Circuit Threshold in Terms of Logic-levels.

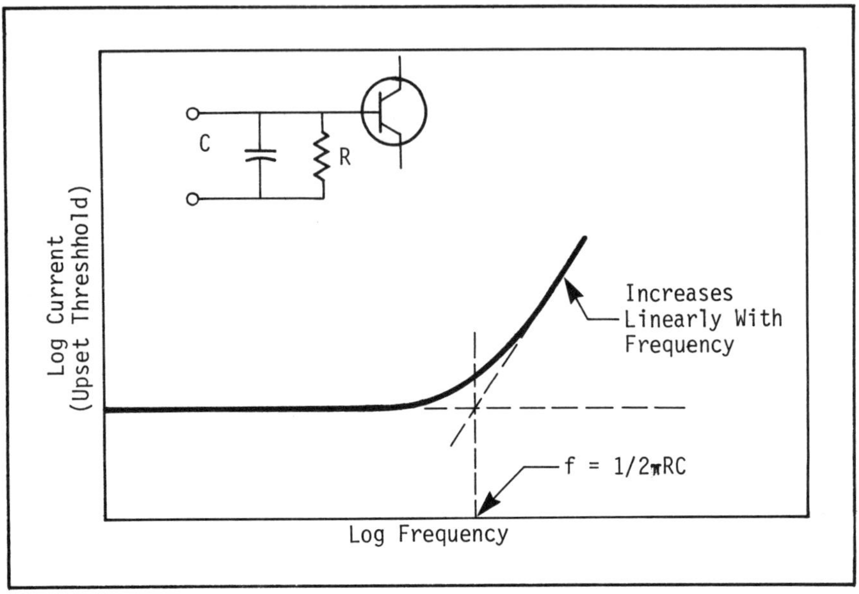

Figure 4.2 - Effect of Shunt Capacitance at the Transistor Input on EMP Threshold. (C Includes Transistor Junction Capacitance).

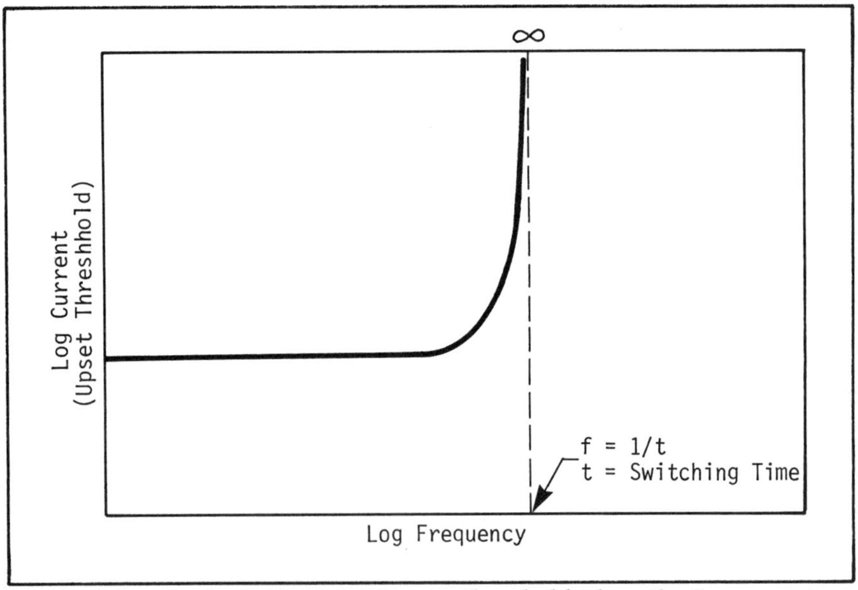

Figure 4.3 - Typical Digital Circuit Threshold when the Response is Limited by Switching Time.

It is sometimes erroneously assumed that a circuit will have its greatest sensitivity or lowest threshold at dc or zero frequency. This is only possible if the circuit is not ac-coupled. Where ac-coupling exists, the circuit threshold will increase at low frequencies as shown in Figure 4.4.

Although experimental data on circuit upsets due to EMP transients for various types of analog and digital circuits is limited, some studies*,**,*** have been made to determine experimentally the degree of circuit upset in logic circuits, pulse-to-dc converters, delay circuits, flip-flops, magnetic gates, and static power switches.

One such study involved the characterization of circuit upset of a NAND**** gate-integrated circuit, where the circuit upset is defined to occur if the output temporarily changes its logical value when a transient appears at some primary input, all other inputs being fixed at logical one. Figures 4.6 and

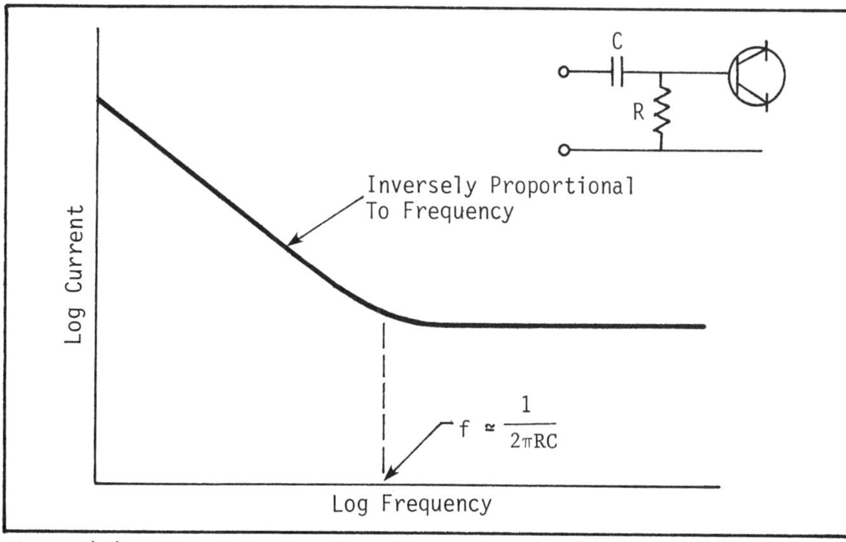

Figure 4.4 - Effect Series Capacitance Between Source and Circuit.

* "Electromagnetic Pulse Handbook for Missiles and Aircraft in Flight," EMP Interaction 1-1, Air Force Weapons Laboratory, September 1972

** R. Goodspeed and T. Lange, "First Pipeline Report on Circuit Thresholds," Boeing Pipeline Report PR-2, October 1969

*** T. J. Lange and B. C. Tappan, "Second Pipeline Report on Circuit Thresholds," Boeing Pipeline Report PR-17, Aug. 1970

****NOT - AND gate.

4.7 show the circuit upset characteristics. Here a positive transient upset is defined as one in which the output change is initiated by a transient input change from a logical zero to logical one. Similarly, if the output change is initiated by a transient input change from logical one to logical zero, it is defined as a negative transient. In general, the gate sensitivity to positive transient upsets differ from the gate sensitivity to negative ones. It is interesting to note in this connection, the following conclusions drawn from the study of input circuit for Sandia Laboratories Diode Transistor Logic Gate as shown in Figure 4.5:

- There is a small amount of voltage amplitude difference between the largest signals which have no probability of causing an upset and the smallest signals which will certainly cause an upset.
- For transients that exceed the threshold for times less than the specified propagation delay, a higher level can be tolerated before transient upset occurs.
- The transient upset is almost independent of the transient waveform and depends largely on the peak value of the transient and the time for which the transient exceeds the dc threshold.
- Input leads or signal reference leads are generally the most susceptible to transient upset.

Figures 4.6 and 4.7 show the upset characteristics of input leads of circuit shown in Figure 4.5.

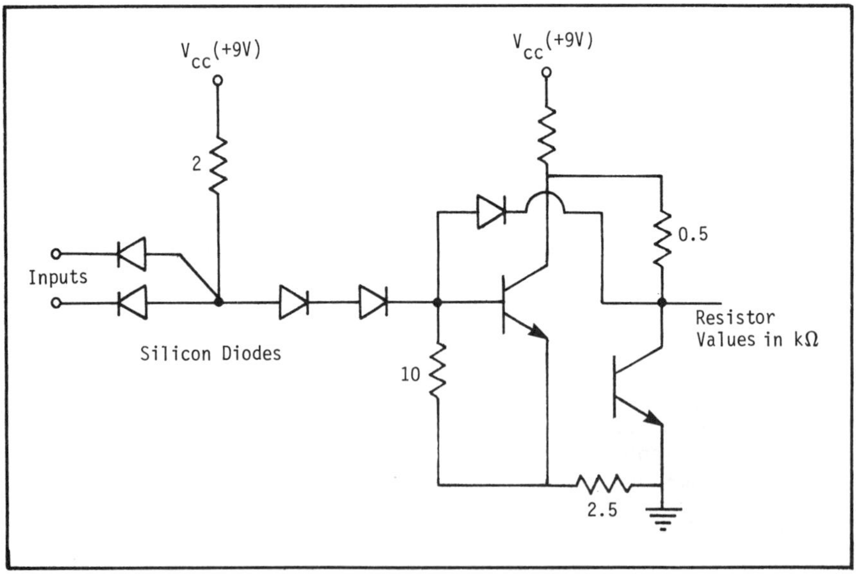

Figure 4.5 - Input Circuit for Sandia Laboratories Diode Transistor Logic Gate.

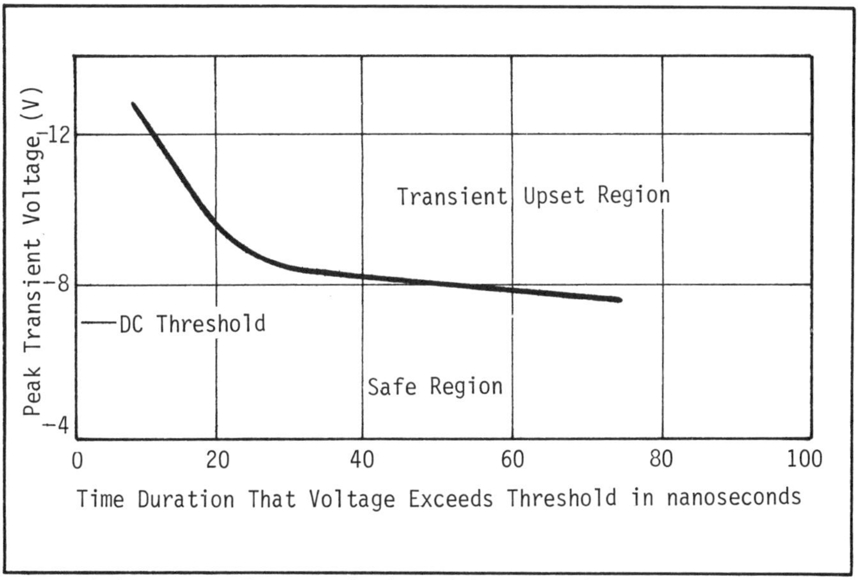

Figure 4.6 - Upset Characteristics of Input Lead of Diode Transistor Logic Gate, (Negative Upset).

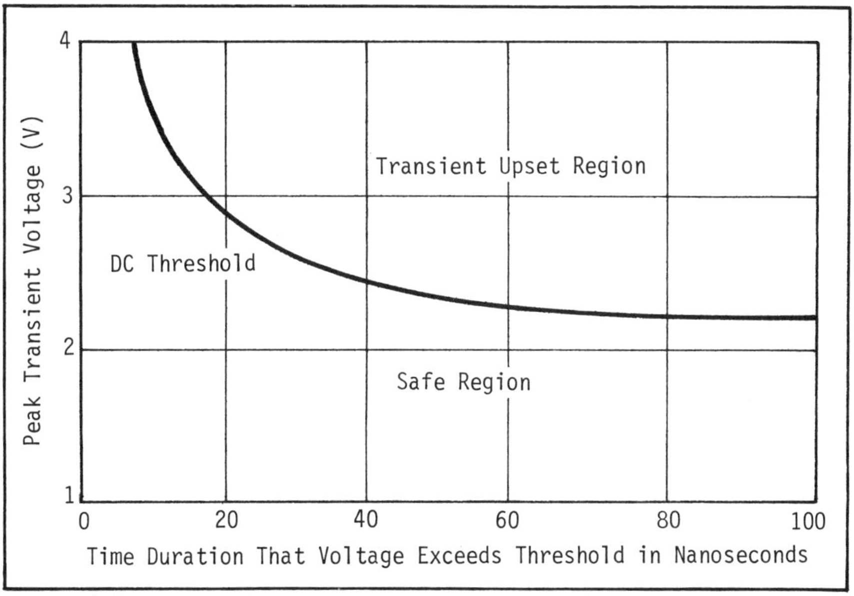

Figure 4.7 - Upset Characteristics of Input Lead of Diode Transistor Logic Gate. (Positive Upset).

Similar generalizations on all other logic or digital circuits which may be susceptible to upset due to EMP transients are not readily available. The frequency dependence of the threshold (voltage) for an "AND" gate, flip-flop circuit and "AND Inverter" are shown in Figures 4.8, 4.9 and 4.10. The threshold voltages shown in these figures are measured values of signals which cause circuit upsets when signals are injected on the wires or terminals, the upset criteria being that the output of the stage is large enough to cause the subsequent stage to switch. The circuits for which the data is presented are not necessarily typical, but the data do illustrate the characteristics of the expected threshold for the types of circuits chosen for the figures. From the data presented in Figures 4.8, 4.9 and 4.10, it is seen that most circuits show an increase in threshold toward the upper end of the frequency band tested. Also, the threshold depends on the input state of the circuit.

The test data in Figures 4.8 to 4.10 are for the sinusoidal waveform of the injected signal, and the threshold seems to show variation with the waveform change of the injected signal.

The threshold signal which may cause an upset for an arbitrary circuit can be estimated from the analysis of the circuit. Examples of such circuit analysis are discussed later in this chapter. Sometimes standard handbook data can be used to determine transient upset probability. For example, a square-pulse triggering voltage is sometimes given as a function of pulse

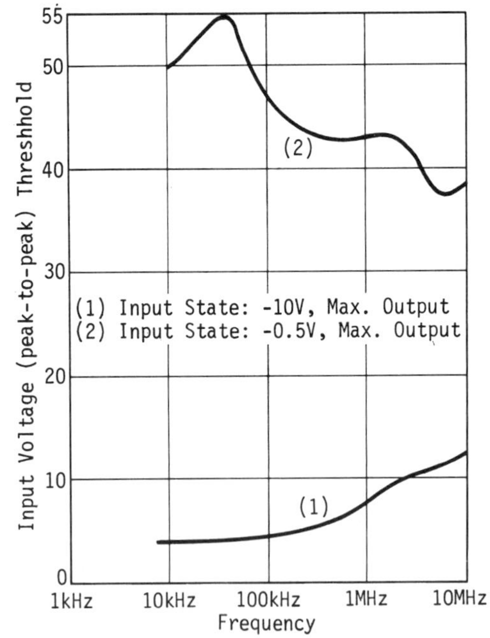

Figure 4.8 - Frequency Dependence of Input Threshold for an"AND" Gate Circuit.

duration to indicate transient upset probability. A typical plot for low-level integrated circuit is shown in Figure 4.11.

It may be noted that the amount of energy required to cause circuit upset or cause erasure of the memory core is very small, on the order of 10^{-9} to 10^{-8} joule.

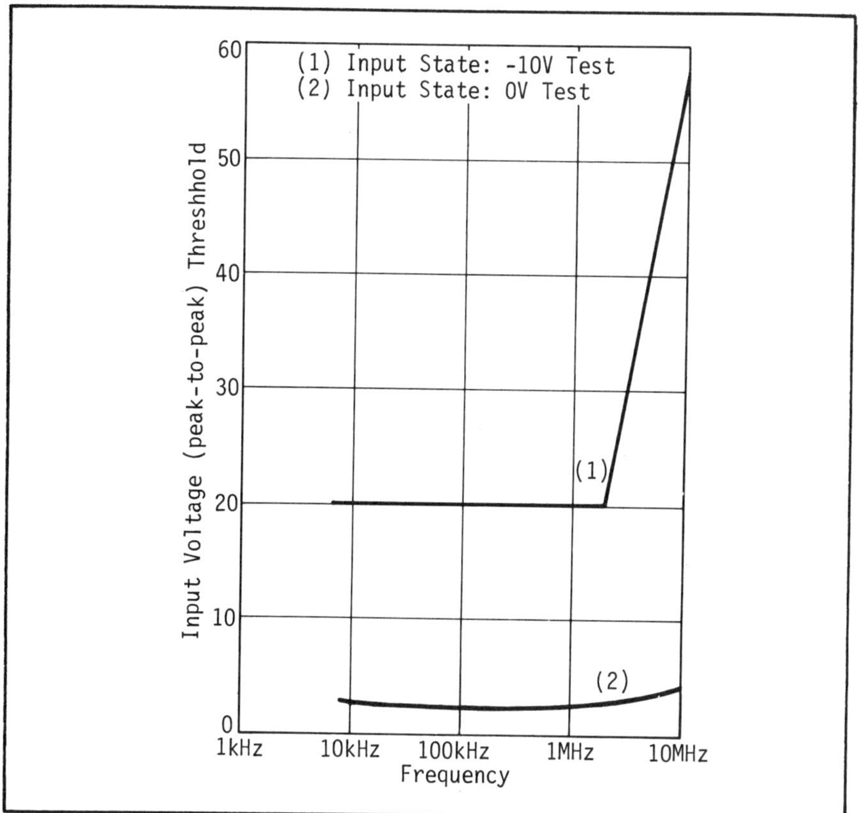

Figure 4.9 - Frequency Dependence of Input Threshold for a Reset Terminal of Flip Flop Circuit.

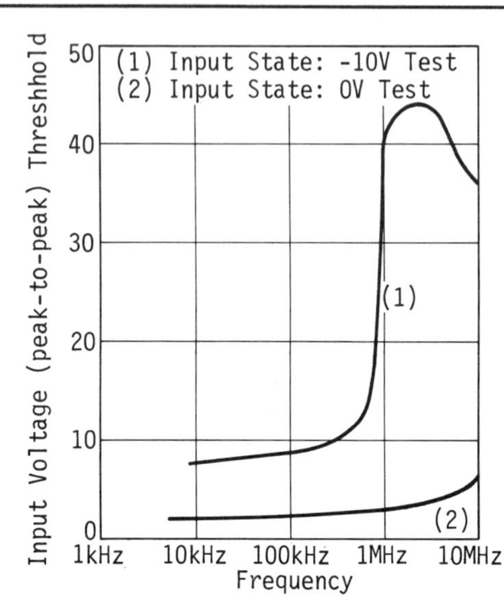

Figure 4.10 - Frequency Dependence of Input Threshold for "AND INVERTER" Circuit.

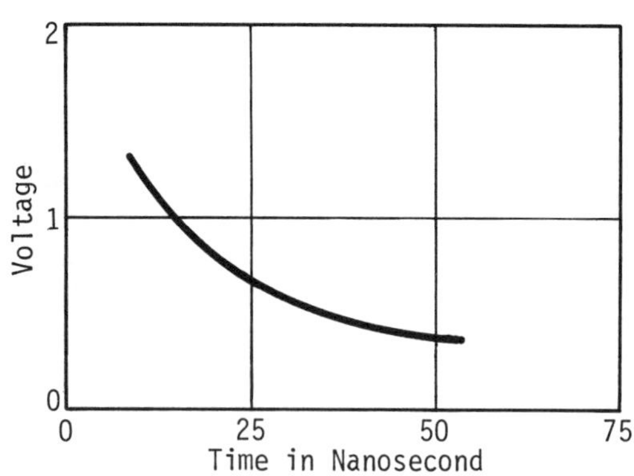

Figure 4.11 - Square Pulse Triggering Voltage vs. Pulse Duration to Indicate Transient Upset Probability.

4.2 Latch-Up

The latch-up is distinguished from the circuit upset discussed earlier, since in latch-up the circuit is not automatically restored, and the power into the circuit has to be removed for the restoration of the circuit. A latch-up due to an EMP-induced transient can occur when the transients flowing through the circuit cause a relay or switch to latch up. A latch-up can also occur within the semiconductor. For example, the n-p-n-p or silicon-control-rectifier can be latched into conductance by an EMP-induced transient, and the power into the circuit has to be removed to unlatch. Latching can also occur in an open-base transistor circuit when the slope of collector current I_c, plotted against V_{ce}, the voltage between the collector and emitter, changes direction beyond some given value of V_{ce} as shown in Figure 4.12. The load line for a particular collector resistance is also shown in the figure. The collector current for such a circuit is normally low (operating point a). However, EMP-induced transient can move the operating level to point b, when the circuit becomes latched up at a high current level. The signal required to cause this event can be determined by noting that the V_{ce} must be driven above the collector-emitter breakdown voltage, V_{ceo}. It should be noted that I_c-V_{ce} characteristics of many transistor circuits are different from what is shown in Figure 4.12, since there is no reversal of the slope of I_c vs. V_{ce} for such circuits.

The possibility of a latch-up of a circuit will depend on the type of circuit under consideration, and certainly many circuits do not latch up. The susceptibility due to latching, therefore, may not exist in many cases. For cases where the possibility of latch-up exists, one has to trace the path of the EMP-induced transient and determine the threshold signal for the latch-up either by an examination of the circuit or by a circuit analysis.

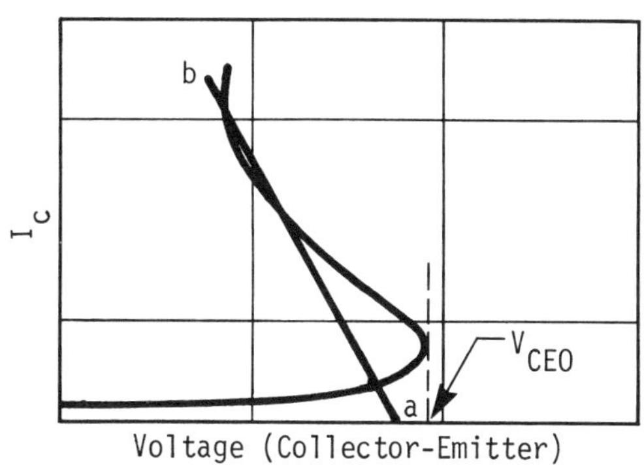

Figure 4.12 - Collector Current vs. Collector-Emitter Voltage to Illustrate Latch Up.

4.3 Permanent Damages

In addition to temporary circuit upsets and latching, as discussed in previous sections, an electronic system can be affected almost permanently by EMP coupled energy, particularly when such an energy causes components burnout. There are various system components which could be susceptible to burnout in an EMP environment, the most sensitive of which are the semiconductors. EMP-induced burnout can, of course, occur also for other circuit components, such as resistors, capacitors, inductors, transformers, relay coils, vacuum tubes, etc.

The likely damages for semiconductors in an EMP environment are junction failures due to the excessive heating of the junction. The primary cause of failures of junctions is believed to be the melting of the silicon in the junction due to high temperature. Furthermore, the junction temperature is not uniform, because the current density is not uniform. This leads to hot spots at the junction and hence weak points to cause a junction failure.

In semiconductors, the damages in an EMP environment are believed to be caused most often by avalanche in the reverse breakdown direction, although the power required to cause junction failure in the forward direction is usually not much higher. For integrated circuits, the failure due to EMP could result from metal interconnection, oxide failure and latch-up of collector-to-substrate junctions.

Wunsch and Bell* developed a theoretical model for the junction failure due to the temperature rise, for one-dimensional heat flow corresponding to a plane junction in an infinite medium.

Based on such a model for the reverse diode current and reverse base-emitter current, and a stepped-pulse type current input, a theoretical expression for the power per unit area of a junction required to damage such a junction is given by:

$$\frac{P}{A} = \overline{K}\, t^{-1/2} \tag{4.1}$$

where $\overline{K}$ is a constant function of time and t is time in seconds. In Equation 4.1, $\overline{K}$ depends on the thermal conductivity k, density ρ, and specific heat C_p of the junction material. It also depends on the difference between the junction failure temperature T_m and the initial temperature T_i. Thus,

$$\overline{K} = (\pi k \rho C_p)^{1/2} (T_m - T_i) \tag{4.2}$$

---

* D. C. Wunsch and R. R. Bell, "Determination of Threshold Failure Levels of Semiconductor Diodes and Transistors Due to Pulse Voltages", IEEE Transactions on Nuclear Science, Vol. NS-15, No. 6, Dec. 1968, pp 244-259.

Assuming typical values of the parameters given in Equation 4.2 as k = 0.526 watts/cm-^{0}K, p = 2.33 gms/cm^3, C_p = 0.7566 joule/gm-^{0}K, T_m = 625^{0}C (898 ^{0}K) and T_i = 27 ^{0}C (300 ^{0}K), one may write from Equations 4.1 and 4.2:

$$\frac{P}{A} = 1109\ t^{-1/2} \qquad (4.3)$$

Experimental data* on diodes and transistors show somewhat lower values of $\overline{K}$ than 1109. They are, for example,

$$\frac{P}{A} = 560\ t^{-1/2} \text{ for diodes}$$
$$(4.4)$$
$$\frac{P}{A} = 310\ t^{-1/2} \text{ for transistors}$$

Although the expressions for P/A as given in Equation 4.4 are not accurate, they are probably valid within less than an order of magnitude. This is illustrated in Figure 4.13, where the power per unit area in kilowatts/cm ^{2} corresponding to burnout is shown, as a function of time, for several transistors.

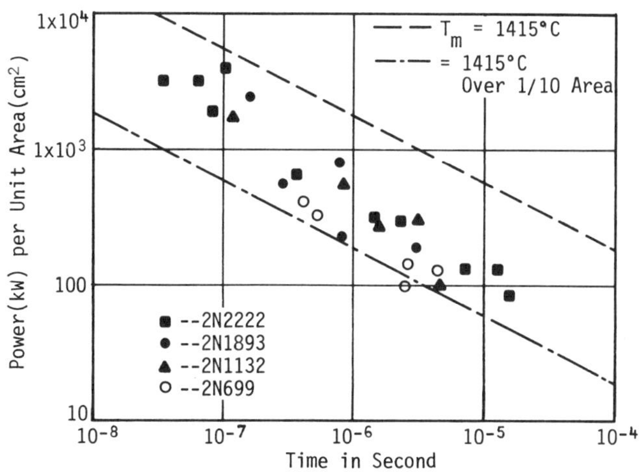

Figure 4.13 - Power per Unit Area Corresponding to Burnout for Several Transistors.

* The discrepancy between the theoretical and experimental values of $\overline{K}$ may be due to the presence of hot spots at junctions, as referred to earlier.

A considerable amount of experimental data on the power required to cause burnout damage to various diodes and transistors are now available. Expressing Equation 4.1 as:

$$P = \bar{K} A t^{-1/2} = C t^{-1/2} \qquad (4.5)$$

experimentally-determined density function of C for a wide variety of diodes, ranging from the very sensitive point contact diode, such as IN82A to the powerful IN2824, is shown in Figure 4.14. A similar distribution of C for various transistors, ranging from the small 2N917 to the large 2N1722, is shown in Figure 4.15. A more detailed breakdown of the ranges of C for various semiconductor-device families are shown in Figure 4.16.

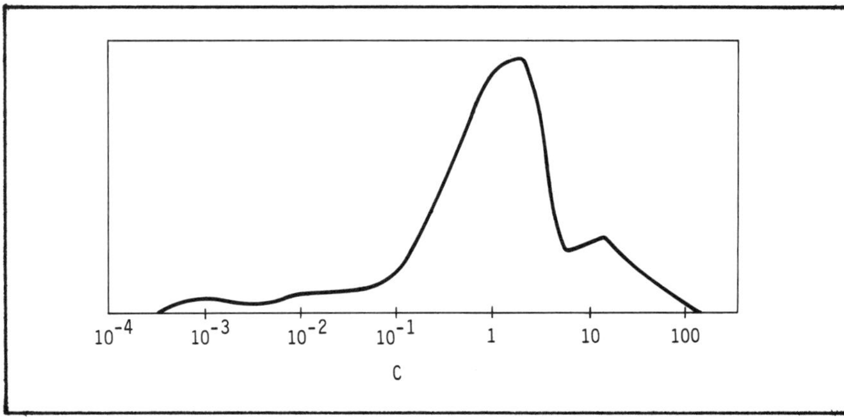

Figure 4.14 - Experimentally Determined Density Function for a Wide Variety of Diodes.

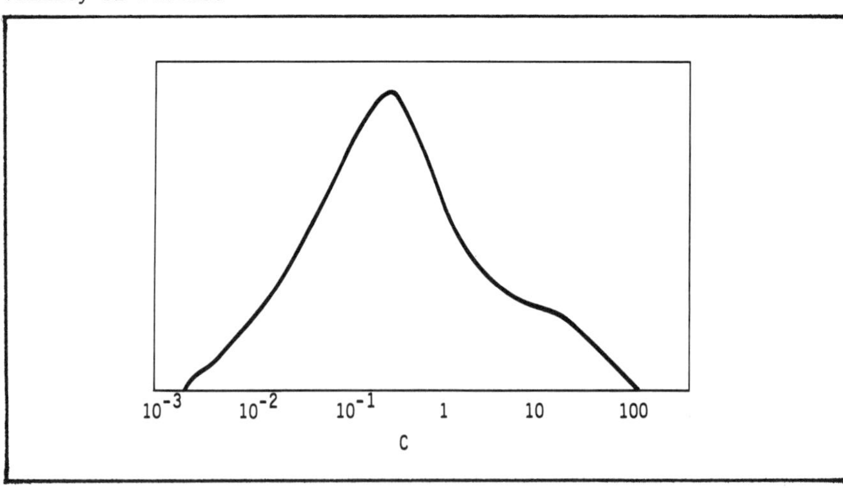

Figure 4.15 - Distribution of C for Various Transistors.

4.3.1 Dependence of Waveform and Pulse Duration on Burnout

The theoretical expression of P/A as given in Equation 4.1 is based on a stepped-pulse type current input to the semiconductor devices. Since the EMP-induced current input is not likely to be of the waveform assumed for the theoretical model, it is desirable to investigate the effect of other waveforms of input current on the burnout P/A. Experimental studies* conducted in this regard, however, lead to the conclusion that the exact waveform of the damaging input current is not nearly as important as the general strength of the semiconductor device. Thus, the burnout data based on the stepped-pulse input current, as discussed in this section, is quite useful in determining the expected behavior of the semiconductor devices in an EMP environment.

The time-dependence of the damaging power per unit area, as given in Equation 4.1, is valid when the EMP energy is applied to the circuit and rings down in about 10 nanoseconds to 10 microseconds. For much larger time duration (greater than 100 microseconds), the power level becomes time-independent, instead of a function of $t^{-\frac{1}{2}}$. For much shorter time duration (less than 10 nanoseconds) of the damaging current input, the burnout power level becomes inversely proportional to time.

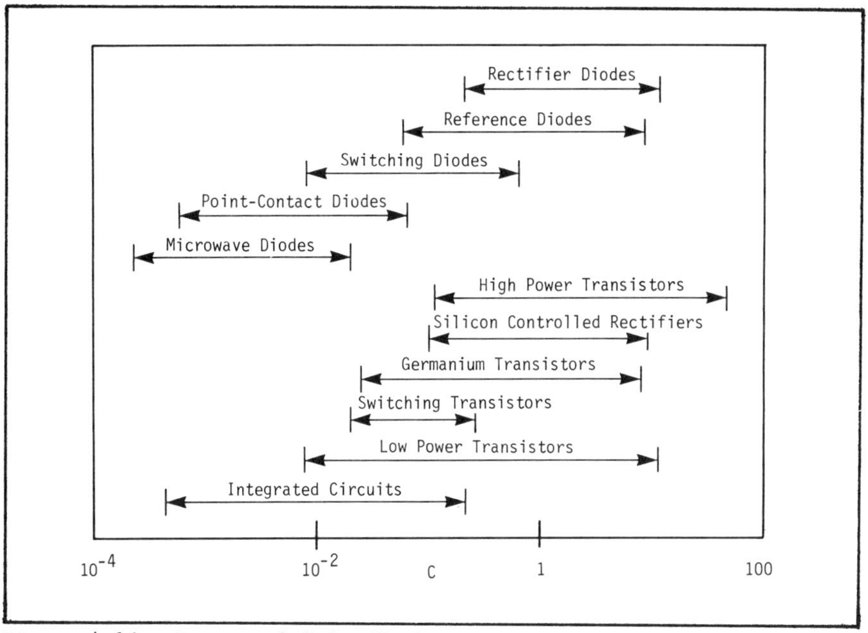

Figure 4.16 - Ranges of C for Various Semiconductor Devices.

---

* D. C. Wunsch and L. Marzitelli, "Semiconductor and Non-Semiconductor Damage Study," BDM Final Report, Vol. I, April 1969.

To estimate the required burnout energy, one can integrate P for the duration of the pulse. For a square pulse of duration t:

$$E = Pt \tag{4.6}$$

An energy representation of the failure threshold for different duration of EMP is shown in Figure 4.17. As seen in the figure, E is a constant of 5×10^{-2} milijoules for $t \leq T_O$, where T_O is between 10 nsec to 1 μsec. In the interval $T_O \leq t < 10$ μsec, E is 10^{-1} milijoule x $t^{-\frac{1}{2}}$, t being in μsec. For $t > 100$ μs, E is 10^{-2} milijoule x t^{-1}.

In addition to junction failures due to elevated temperature, another principal breakdown mode for semiconductor PN junctions under applied reverse-bias voltages is the surface damage around the junction due to arcing. Such a damage refers to the establishment of a high leakage-path around the junction which effectively eliminates any junction action. The junction itself, in this case, is not necessarily destroyed since etching the leakage-path conducting material from the surface may return the device to its normal operating state. However, if the formation of such a surface leakage-path is the result of excessive heat production in the bulk material, the failure mechanism is likely to be the result of some bulk phenomenon and the normal

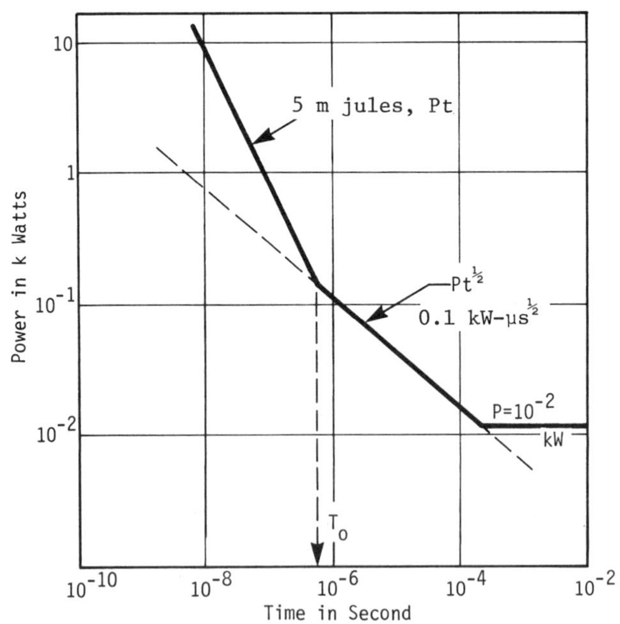

Figure 4.17 - Energy Representation of the Failure Threshold for Different Duration of EMP.

operating state of the device cannot be recovered. It is difficult to predict theoretically the conditions which will lead to surface damage since they depend on many variables, such as the geometrical design and the details of the crystal structure at the surface. It is believed, however, that a surface damage may occur in devices at power levels which are orders of magnitude below that sustainable by devices in which the bulk damage occurs.

As mentioned earlier, the heat energy resulting from EMP can cause a permanent damage to circuit components other than semiconductors. The energy required to cause such a damage will depend on the type of materials used in components and the nature of the resulting malfunction. A damage in the form-welded contact in a relay, for example, may require 0.1 to 0.0001 joule of energy. Similarly, a damage in the form of a slammed meter for a microammeter may require 1 to 3 x 10^{-3} joule. Explosive bolts, squibs, etc., may require 10^{-3} to 10^{-4} joule of energy for the ignition-type damage.

Another source of damage to non-semiconductors, such as capacitors is due to the excessive voltage rise across their terminals instead of heat. When the EMP-induced voltage across the tantalum electrolytic type capacitors exceeds 3 to 4 times their voltage ratings, for example, these components begin to conduct. Initially, the leakage current increases monotorically with the time of conduction. The value of the capacitance usually determines how quickly the voltage across the capacitor reaches the breakdown range.

The susceptibility of an electronic system due to nuclear EMP, either in the form of temporary circuit upset, latch-up or permanent damage, will depend on the coupling of EMP energy in various circuits and their components. An assessment or prediction of susceptibility due to EMP can be made by appropriate circuit analysis. Alternatively, one can also assess susceptibility by employing a transfer-function technique, where the amplitudes and phases of voltages or currents or both, due to a known low-level CW source at various frequencies, are monitored at various sensitive points of concern in the circuits. The equivalent transfer functions between the source and each sensitive point in the circuits can then be used to extrapolate the expected magnitudes of the time-varying voltages and currents for any given level and waveform of the EMP threat field affecting such circuits. These techniques of susceptibility assessment and prediction are further discussed in the following sections.

4.4 Transfer Function Technique to Assess Susceptibility

In general, the susceptiblity of an electronic system due to nuclear EMP will depend on the level of voltages or currents at the inputs of sensitive components of the system or the energies delivered to such components. For a known EMP field as the source, the input voltage or current at any sensitive component, or the energy delivered to the component, can be determined by a circuit analysis method. When the circuits containing the sensitive components are very complex, one may need the aid of an electronic computer for determining the voltage or current level referred to above. Sometimes, the circuits in a large electronic system can be too numerous and too involved, because of their interconnections, to permit a straightforward circuit analysis even with a computer. An alternative and perhaps more parctical approach for such a case is to combine analytical and experimental techniques for the solution of the problem. The use of the transfer-function technique appears to be one such alternative.

The concept of the transfer-function technique can, perhaps, be best illustrated with Figure 4.18. Here $E(\omega)$ denotes the source field and $V_i(\omega)$ represents the voltage at an arbitrary point i of a network as a function of the angular frequency ω of the source. Let the transfer function between the source and the point i of the network be denoted as $T_i(\omega)$, such that:

$$E(\omega)\, T_i(\omega) = V_i(\omega) \tag{4.7}$$

In a linear medium or network, $T_i(\omega)$ is independent of $E(\omega)$. Thus, if $E_s(\omega)$ and $V_{is}(\omega)$ denote respectively a test source and the voltage at the ith point due to this test source, the voltage $V_i(\omega)$ at the same ith point due to another source $E(\omega)$ becomes:

$$V_i(\omega) = E(\omega)\, V_{is}(\omega)/E_s(\omega) \tag{4.8}$$

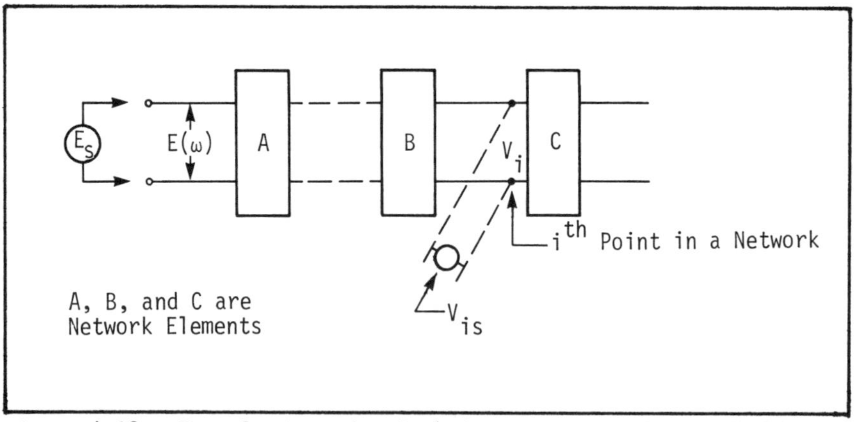

Figure 4.18 - Transfer Function Technique to Assess Susceptibility.

For the assumed notation of the Fourier-transform pair:

$$F(\omega) = \int_{-\infty}^{\infty} f(t)\, e^{-i\omega t} dt \tag{4.9}$$

and

$$f(t) = \frac{1}{2\pi} \int_{-\infty}^{\infty} F(\omega)\, e^{i\omega t} d\omega$$

one can write the time-varying response of $V_i(\omega)$ in Equation 4.8 as:

$$V_i(t) = \frac{1}{2\pi} \int_{-\infty}^{\infty} E(\omega)\, \frac{V_{is}(\omega)}{E_s(\omega)}\, e^{i\omega t}\, d\omega$$

$$= \frac{1}{2\pi} \int_{-\infty}^{\infty} E(\omega)\, T_i(\omega)\, e^{i\omega t}\, d\omega \tag{4.10}$$

It should be noted that $E(\omega)$ and $T_i(\omega)$ in Equation 4.10 are complex variables of the real variable ω and therefore, to determine $V_i(t)$, one must not only know the magnitude of $E(\omega)$ and $T_i(\omega)$ but also their corresponding phases.

To illustrate the applicability of the transfer-function technique, let it be assumed that the mode of entry of EMP energy in a system is an antenna, and i, j, and k are sensitive points in the system where component malfunction or burnout may occur due to EMP. If one injects a CW signal of known voltage $V(\omega)$, at angular frequency ω, across the antenna terminals and measures the voltages across the sensitive components at points i, j, and k, one may write the transfer functions corresponding to these points as:

$$T_i(\omega) = V(\omega)/V_i(\omega)$$

$$T_j(\omega) = V(\omega)/V_j(\omega)$$

$$T_k(\omega) = V(\omega)/V_k(\omega).$$

where $V_i(\omega)$, $V_j(\omega)$ and $V_k(\omega)$ are voltages at points i, j, and k as referred to above. As mentioned earlier, $T_i(\omega)$, $T_j(\omega)$ etc., will be complex quantities having magnitudes and phase angles. If now one determines $T_i(\omega)$, $T_j(\omega)$, etc., for all expected values of ω in the EMP at the antenna terminals, one can write the time-varying voltages $V_i(t)$, $V_j(t)$ and $V_k(t)$ at points, i, j and k as:

$$V_i(t) = \frac{1}{2\pi} \int_{-\infty}^{\infty} E(\omega)\, T_i(\omega)\, e^{i\omega t}\, d\omega$$

$$V_j(t) = \frac{1}{2\pi} \int_{-\infty}^{\infty} E(\omega)\, T_j(\omega)\, e^{i\omega t}\, d\omega \qquad (4.11)$$

$$V_k(t) = \frac{1}{2\pi} \int_{-\infty}^{\infty} E(\omega)\, T_k(\omega)\, e^{i\omega t}\, d\omega$$

where $E(\omega)$ is the Fourier transform of the voltage $e(t)$, due to EMP, at the antenna terminals, that is,

$$E(\omega) = \int_{-\infty}^{\infty} e(t)\, e^{-i\omega t}\, dt \qquad (4.12)$$

Once the voltages $V_i(t)$, $V_j(t)$, etc., are known, the susceptibility of the system can be assessed based on the sensitivities of components at points, i, j, etc. Equation 4.11 utilizes both analytical and experimental techniques for any EMP threat, the key elements of the equation being $T_i(\omega)$, $T_j(\omega)$, etc., which are determined exclusively from experiments.

Principal advantages of the transfer-function technique are:

- Characteristics of the circuits in a system are determined, particularly for sensitive points, with the system installed in place taking into account any mutual interactions among circuits. It may not always be possible to account for such interactions in the circuits analysis approach to be discussed in the following section.

- Since transfer functions are determined experimentally with a CW source signal, one can employ an extremely narrowband receiver (phase-lock loop), thereby requiring only modest power for the signal source.

Principal disadvantages of the transfer-function technique are:

- The technique is not suitable for circuits involving nonlinear elements where Fourier-transform technique for circuit analysis cannot be used.

- Experimental determination of transfer functions are tedious and time consuming.

- Errors in transfer-function determination may result unless adequate data points (for a large number of values of ω) are taken to represent the continuous functions $T_i(\omega)$, $T_j(\omega)$, etc.

However, the transfer-function technique may be the only technique available for susceptibility assessment in many cases, particularly when the system is very involved, and adequate information on some parts of the circuits or the system is not available for the detailed circuit analysis.

4.5 Circuit Analysis

The goal of the susceptiblity analysis of a system involving electronic components is to ascertain whether or not one or more components in the system are expected to fail under the environmental stress resulting from EMP. This may be accomplished in two steps. The first step constitutes the estimation of the signal levels at the terminals of sensitive components due to EMP. The comparison of these signal levels to the estimated or measured threshold-damage levels of the components is the second step. As mentioned in the previous section, one approach toward achieving the first step is to utilize the transfer function technique, which combines experimental and analytical means to determine the signal levels at the sensitive point in a system. Another approach to achieve the same goal is the circuit analysis. The circuit analysis for EMP susceptibility assessment can be hand analysis or it can be computer-aided.

An electronic system contains one or more circuits. A complex system usually contains many circuits, often interconnected. Also, not all circuits have the same or similar impact on the mission of the system. Therefore, the damages, failures, or malfunctions of components in all circuits may not be critical from the viewpoint of the system's mission or missions. To avoid some unnecessary and time-consuming efforts toward the analysis of all circuits, it is necessary to:

- Subdivide the system into subsystems and rate the subsystems according to the mission criticality.

- Assign priorities among critical subsystems, based on the degree to which the damage or failure or malfunction of components in such subsystems would impair significantly, or nullify, the capability of the complete system to perform its defined mission.

- Survey all mission critical input/output circuits in a critical subsystem and select those which contain unprotected sensitive components.

- Determine the damage and/or upset threshold of the sensitive components. For example, V_{bd} (breakdown voltage) for solid state devices and capacitors or the damage constant $\overline{K}$ for solid state devices.

Before performing any circuit analysis one needs to examine the circuits in the highest priority subsystems to determine whether they contain solid state devices or capacitors not protected by resistive or impedance type circuit elements and whether they are directly coupled to wiring which provides access to inter-compartment cabling. These criteria are selected to identify circuits which have a high probability of either containing damage-prone devices or of being exposed to relatively large transient EMP-induced signals. The second criterion is not applicable where upset is a possible damage mechanism. Only those circuits which meet the criteria referred to above need to be analyzed further to determine their potential EMP susceptibility.

The consideration of the waveform of EMP is also necessary before performing the circuit analysis. EMP, as the name implies, is in the form of a pulse. The short rise-time of such a pulse is potentially more damaging for a given pulse width than a long rise-time. This is because fast rise-time pulses may allow a higher voltage to be developed across a PN junction for a longer time. To a first approximation, the most damaging portion of the EMP signal can be approximated by a square pulse which has an amplitude equal to the maximum voltage.

It should be noted that the approximation of the square pulse should be made for the equivalent EMP of concern at the circuit and not the free-space EMP signal arriving at the system. Thus the characteristics of the EMP at the output of the antenna or cables of the sytem, or those following the penetration through shields or apertures, should be the basis of the square pulse approximation.

The circuit analysis can, of course, be carried out for a CW signal, and Fourier-transform technique can then be used to predict the final results for any given waveform of the EMP.

When the circuits of concern are simple, a hand analysis for the susceptibility assessment will be adequate, and no computer aid will be required. For example, suppose it is required to calculate the input power level required to destroy the diode in a circuit shown in Figure 4.19 due to an EMP. If a damage constant, $\overline{K}A = 0.219$ (w.sec$^{\frac{1}{2}}$) is assumed, the power required for the junction in the diode to fail for a 10 μsec rectangular pulse is:

$$P_{diode} = \overline{K}At^{\frac{1}{2}} = (0.219)(10 \times 10^{-6})^{-\frac{1}{2}} = 69.25 \text{ watts} \qquad (4.13)$$

If a breakdown is assumed at this point, the current through the junction can be calculated from the relation:

$$I = \frac{P}{V_{bd}} \tag{4.14}$$

V_{bd} being the breakdown voltage. If V_{bd} is equal to 5.8 volts, the current through the resistance R_2 is:

$$I_2 = \frac{69.25}{5.8} = 11.94 \text{ amps}$$

The voltage drop across R_2 and the power dissipated in R_2 can be written as:

$$V_{R_2} = 11.94 \times 500 = 5.97 \text{ K volts}$$

(4.15)

$$P_{R_2} = (11.94)^2 \times 500 = 71.38 \text{ K watts}$$

The voltage across the capacitor C_1:

$$= V_{C_1} = 5970 + 5.8 \simeq 5.976 \text{ K volts}$$

Input voltage required to charge C_1 so as to develop 5.976 K volts is:

$$V_o = \frac{V_{C_1}}{1 - e^{-t/R_1C_1}}$$

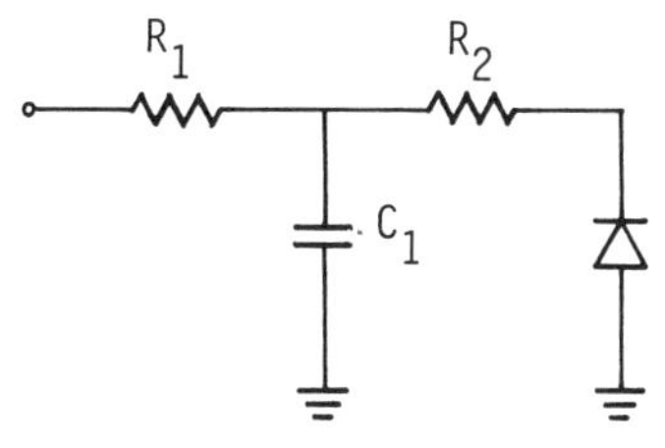

Figure 4.19 - A Simple Circuit Containing a Semiconducting Device for Hand Analysis.

When $R_1C_1 = 10^{-5}$, and t = 10 μ sec, the power dissipated in R_1 is given by:

$$P_{R_1} = (V_o - V_{C_1})^2/R_1 = 11.9 \text{ K watts} \quad (4.16)$$

The power stored within the capacitor is:

$$P_{C_1} = \frac{V_C{}^2 C_1}{2t} = \frac{(5.976)^2 \times 0.1 \times 10^{-6}}{2 \times (10 \times 10^{-6})} = 17.8 \text{ K watts} \quad (4.17)$$

The total input power required for the diode to fail is:

$$P_{total} = \text{Sum of power within the circuit}$$

$$= P_{R_1} + P_{C_1} + P_{R_2} + P_{diode} = 101 \text{ K watts} \quad (4.18)$$

The above analysis of an idealized circuit shows that a much larger EMP signal power than the diode breakdown power is needed at the circuit input to cause the failure of the diode. It also shows that perhaps the more vulnerable elements in the circuit under consideration are the resistor or capacitor rather than the diode.

Another example of a circuit which is amenable to hand analysis is shown in Figure 4.20. Here $R_1 = R_2 = 100$ ohms, $C_1 = 100$ p farad, L_1 is large, Q_1 = a transistor with K = 0.1 watt-sec$^{1/2}$ and Breakdown Voltage $V_{bd} = 5V$. Let the input pulse width be 10^{-8} sec.

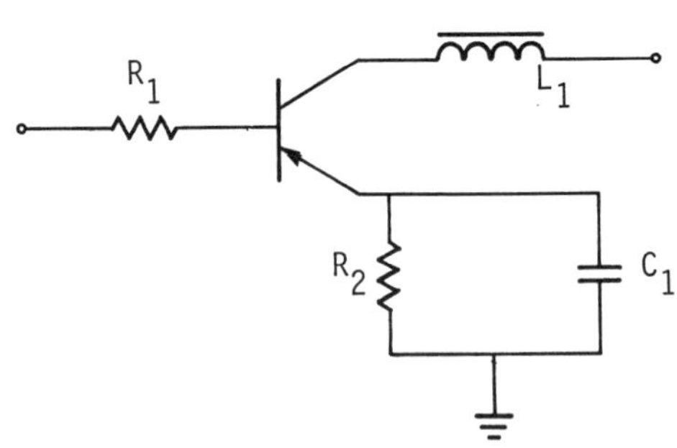

Figure 4.20 - A Simple Circuit Containing a Transistor Suitable for Hand Analysis.

The power dissipation at the base-emitter junction of Q which is required to cause its failure is:

$$P = Kt^{-1/2} = 0.1\,(10^{-8})^{-1/2} = 1\text{ K watt} \tag{4.19}$$

If the base-emitter breakdown is assumed at this point, the current through the junction can be calculated from the relationship:

$$I = \frac{P}{V_{bd}} = \frac{1000}{5} = 200\text{ amps}$$

Since L_1 is large, this 200 A must all come from the input, so that the input current, I_{input} = 200 amps. But if 200 amperes are flowing into the R_2 - C_1 parallel circuit, the voltage at the emitter of Q_1 becomes:

$$V_{C_{Q_1}} = I_{in}\,R_2\,\left(1 - e^{-(t/R_2C_1)}\right)$$

$$= 20000\,\left(1 - e^{-t/10^{-8}}\right)\text{ volts}$$

Since the same current is flowing through R_1, the voltage across R_1 is:

$$V_{R_1} = I_{input} \cdot R_1 = 20\text{ K volts}$$

The voltage at the input can now be found by adding the voltages from the ground to the input. Thus,

$$V_{input} = V_{C_{Q_1}} + V_B + V_{R_1}$$

$$= 20000\,\left(1 - e^{-t/10^{-8}}\right) + 5 + 20000$$

and the input power P_{in} is given by:

$$P_{in} = \frac{1}{T} \int_{0}^{T} (V_{in} \cdot I_{in})\, dt$$

$$= \frac{1}{10^{-8}} \int_{0}^{10^{-8}} \left[\left\{ 20000\,(1 - e^{-t/10^{-8}}) + 5 + 20000 \right\} 200 \right] dt$$

$$\simeq 8 \times 10^{3}\ \text{K watts} \tag{4.20}$$

Here again, it is seen that the required input power is much greater than the junction power to cause a junction failure.

In summary, the steps of the hand analysis of a circuit for the susceptibility analysis may be outlined as follows:

- Use the device-damage constant to find the failure current through the junction.
- Use junction-failure current to find the associated circuit voltages and currents.
- Sum the power dissipated in the circuit components to find the total input power.
- Alternatively, calculate the current I and voltage V as a function of time, and then calculate P_{total} as:

$$P_{total} = \frac{1}{T} \int_{0}^{T} V\, I\, dt \tag{4.21}$$

Unfortunately, the analysis of circuits and electronic subsystems to assess their susceptibilities due to an EMP can become extremely complicated as the number and types of components involved in the circuits are increased. By automating the analysis on a digital computer, one can not only expedite the circuit response calculations for a large number of circuits, but also incorporate a much more detailed treatment of components.

A large number of circuit-analysis computer codes are available and are sufficiently documented to allow their general usage. Among the better-known codes are:

ECAP (IBM)
NEDAP (U. S. Army)
SCEPTRE (IBM)
CIRCUS
NET-2 (Los Alamos)

The purpose for developing these codes was to provide a mathematical tool both for predicting response of a circuit with a given set of components in a given configuration and for use in circuit design. These circuit analyses enable the user to solve even several classes of nonlinear problems arising from:

- nonlinear volt-ampere characteristics of semiconductor junctions.
- variation in active device gain with respect to operating point.
- variation in device junction capacity with respect to current and voltage.
- photocurrent variation with respect to junction voltage and operation current levels, and the variation in carrier life-time and gain.

To describe component characteristics completely, large demands are placed on storage capabilities for all codes. Some of the features of various circuit analysis programs are:

- Basic Circuit Elements, such as resistors, capacitors, etc., are introduced by numbering nodes and branches and giving values to elements connected between the nodes.
- Switch: A switch can be used to change a circuit, or to change circuit behavior, when the proper conditions are met. This allows applying linear analysis to nonlinear problems.
- Controlled Current Sources: These sources allow current in one branch of the circuit to be sensed, multiplied by an arbitrary factor, and introduced into another branch. This feature may be used to model transistor function.
- Diode and Zener Models: Diodes and Zener characteristics can be approximated by straight-line representations with discontinuous slope characteristics.
- Transistor Models: Transistor models usually include cut-off, active region, large signal region, saturation, Zener models, power-

dissipated th junctions, damage prediction, and simulation of the damage.

- Damage Prediction: Power can be specified as limited in the collector-base junction, emitter-base junction, or as total transistor power, diode junction power and general branch power. A power tolerance curve of the form $P = Ct^{-n}$ can sometimes be specified. When this power is exceeded, execution of analysis may be terminated or continued with a damaged model.

- Driving Waveform: An EMP waveform can be entered by modeling straight-line segments of any time duration as a current or voltage source.

As mentioned earlier, there are many good codes for circuit analysis in existence, and each has its own advantages. Some of the considerations that must go into a choice of code can be listed in order of importance as follows:

- Availability (of the particular computer to be used).
- Flexibility to include breakdown models.
- Cost.
- Flexibility, such as the ability to change parameters in the middle of a run, restart capabilities, etc.

The computer-aided circuit analysis is suited for complex and involved circuits which are not easily amenable to hand analysis. Because of the diverse and varied nature of such circuits, it is difficult to select a typical circuit for illustration of computer analysis techniques. Furthermore, such an exercise is beyond the scope of this text, and therefore is not attempted here.

4.6 Susceptibility Prediction for Various EMP Threats—Preliminary Analysis

Notwithstanding various means of circuit analysis including computer aids, as discussed in the previous section, it is often not practical to obtain an accurate determination of all EMP effects on a system. However, the determination of all EMP effects may not even be necessary in most cases. For example, from the preliminary examination of a system, it is often possible to identify a number of subsystems or circuits where the expected rise of voltage or current due to an EMP will have negligible effects, if any, insofar as the performances of such subsystems or circuits are concerned. Obviously, an exact determination of the rise of voltages or currents due to an EMP in such cases will have little usefulness. On the other hand, a similar preliminary examination of the system may show critical areas of a system when a circuit upset or a component burnout may have a severe detrimental effect on the mission of the system. Undoubtedly, a careful investigation of such areas are necessary to assess EMP effects for the expected EMP threats. This type of investigation, involving the determination of EMP-coupled signal into the system and appropriate circuit analyses for EMP-coupled signal thus determined, could be time-consuming and should be kept to a minimum from the cost-effective consideration alone. In many cases, an interim analysis may be undertaken to identify critical problems and areas and to set bounds for such problems. In particular, the interim analysis will be useful for:

- Determining whether a potential EMP problem exists for a given threat and system.
- Isolating critical areas where adverse EMP effects will have intolerable detrimental impact on the given mission of the system.
- Determining shielding effectiveness, filtering rejection or similar requirements on a preliminary basis.
- Estimating waveforms and levels to be used for circuit analysis.
- Assisting the design engineer for a new system (not yet built) in circuit layout, system grouping, cable routing, shielding, and the selection of components.

For the interim analysis, the given EMP environment and the known component susceptibility in each critical subsystem or circuit constitute the starting point. Based on the geometry of the structure, antennas, if any, and related cable runs, one can then develop a plausible worst-case energy coupling mechanism for EMP into the system of concern. Using the environment and coupling, the worst-case collected energy can then be determined. This energy can be compared with the minimum energy required to degrade a component in a critical subsystem or circuit, either functionally or operationally. If the collected energy exceeds the minimum energy required to degrade the component, then a potential problem has been identified. It should be noted that this procedure merely demonstrates the possibility of an event and not its probability.

The worst-case energy analysis may also be used to assist in the identification of critical areas or weak links in a system. The value of this approach does not lie in the quantitative results, but rather in the fact that it aids the system engineer in viewing the entire system from the viewpoint of EMP-induced system degradation.

Following the preliminary analysis, as described above, a more detailed analysis may often be required for the assessment of susceptibility or vulnerability of the system. A methodology for the vulnerability analysis of systems subjected to EMP threats are further discussed in Chapter 7.

4.7 References

1. *Electromagnetic Pulse Handbook for Missiles and Aircraft in Flight EMP Interaction 1-1*, Air Force Weapons Laboratory, September 1972.

2. Gray, R. M., *An Experimental Investigation of EMP-Induced Transient Upset of Integrated Circuits*, SC-TM-71 0330, Sandia Laboratory, July 1971.

3. Wunsch, D. C. and Bell, R. R., *Determination of Threshold Failure Levels of Semiconductor Diodes and Transistors due to Pulse Voltages*, IEEE Transactions on Nuclear Science, Vol. NS-15, No. 6 DCC 1968, pp. 244-259.

4. Ebers, J. J. and Moll, J. L., *Large-Signal Behavior of Junction Transistors*, Proc, IRE, Vol. 42, December 1954, pp. 1761-1772.

5. Daniel, M. E., *Development of Mathematical Models of Semiconductor Devices for Computer Aided Circuit Analysis*, Proc. IEEE, Vol. 55, November 1967, pp. 1913-1920.

6. Hart, W. C., *EMP Computer Code Survey*, DASA 2345, DASA Information and Analysis Center, Santa Barbara, California, August 1969.

7. Davis, K. E., and Keyser, R. C., *Input Circuit Threshold Analysis*, Gulf General Atomic Report No. GACD-9770, October 1969.

8. Bastian, J., *EMP Circuit Analysis Example*, Autonetics IL 71-552-21-EDA-132, July 1971.

9. Johnson, E. D., Kleiner, C. T., et al, *Transient Radiation Analysis by Computer Program*, Vol. 1, Users Guide, AD 304, September 1968.

10. Tasca, D. M., *Power Failure Modes in Semiconductors*, presented at IEEE Annual Conference on Nuclear and Radiation Effects, July 1970.

CHAPTER 5

System Vulnerability Analysis for EMP Threats

EMP vulnerability analysis of a system involves the determination of system response and the assessment of probability of failure of the intended mission of the system in a specified or expected EMP environment. Factors which affect the EMP vulnerability of a system are (1) the environment, (2) system response, (3) system threshold, (4) system mission, and various uncertainties in the environment, system response etc. For the EMP vulnerability analysis, an exact determination of voltages and currents at all points in circuits or the system is usually not necessary. Instead, one needs to formulate a systematic approach for determining EMP-induced inputs to functionally sensitive components or subsystems, which are vital to proper system operation from the viewpoint of the specified system mission, and which may suffer functional degradation due to EMP. Such a systematic approach or methodology is presented in this chapter. The vulnerability analysis is facilitated by system zoning which identifies those volumes of space for which an estimate of EMP-induced susceptibility must be determined. A theoretical approach to assess susceptibility of an arbitrary system is also outlined in this chapter.

5.1 Variables in Vulnerability Analysis

EMP vulnerability of a system may be defined as the probability of failure of the system to perform its intended mission due to EMP. Factors which affect EMP vulnerability are:

- EMP environment.
- System response.
- System threshold and sensitivity.
- System mission.

To obtain a quantitative measure of the vulnerability, one needs to define the above factors quantitatively. Such a task is usually not easy for most systems and for most EMP threats.

The convenient starting point for the vulnerability analysis is to define EMP threat. Normally this means the definition of the time-varying E and H fields at the various energy collector points of the system, such as cables, antennas, etc., due to EMP. Such a definition, however, becomes involved particularly when the system could be subjected to EMP onslaught due to either a high-altitude, low-altitude or surface nuclear burst. It has been seen in Chapter 1 that the characteristics of E and H fields, their magnitudes and waveforms, may differ significantly depending on the altitude of the burst, even when the yield of the weapon producing EMP is the same in each case. Thus, EMP environmental threat may vary depending on its source, frequency band, EMP rise and fall times, intensity, regions of burst, air conductivity induced by ionization, and the nature of the victim system. Sometimes, system susceptibility could result from synergistic effects of multiple threats.

Another important variable in EMP environment is the polarization of the field. Both the polarization of EMP and its direction of propagation vary as a function of the relative position of the burst and the victim system. Since such a relative position will seldom be a priori known, the task of defining the E and H fields will be difficult. Consequently, the task of determining EMP-coupled field or voltages and currents into the victim system, necessary for the quantitative assessment of EMP vulnerability, will be difficult.

The polarization of EMP and its direction of propagation not only affect the coupling of EMP into the system, but also modify the very field of concern, at least for the ground-based systems. Since the resultant field in the presence of the earth's surface will, in general, be modified by ground reflection, which in turn depends on the conductivity and dielectric constant of the local earth and polarization and direction of propagation of EMP field, uncertainties may exist in the definition of E and H fields, even when the weapon yield producing EMP of concern remains the same.

Other variables which may, at times, affect EMP environment include climatic conditions such as rain, snow, temperature, altitude of the victim system, humidity, etc., background ambient noise level, and terrain configuration in the region of weapon burst and the victim system.

System response for any specified EMP field is also not invarient. For aircrafts and missiles, the system response, for example, will be significantly different when the system is in flight from the corresponding response on ground for the same EMP environment. This difference is not only due to the normal difference of performance of most electronics with altitude and pressure, but also due to the plasma that results from the burning gas in case of missiles, different wing shapes for aircrafts in flight, absence of the earth as a reflective surface, etc. The system response could also be different when it is new as compared with the response due to the same EMP environment when the reliability of components, such as switches, relays, etc., has been lessened with usage. Furthermore, the system response could be different from the normal expected response when a routine or statistical failure of any component or components has weakened the system before EMP strikes.

System threshold, another factor affecting the vulnerability analysis, is also not invariant. One may define the system threshold as the level of voltage, current or power input to a system or components of a system which will cause the system to fail in accomplishing its mission. System threshold may change with age and use of the system and with the development of a "weak link" in the system either through natural deterioration during its operating life or at random, based on some statistical probability. Furthermore, the weak link may be temporary or permanent.

From the operating knowledge of a system and from its design, one can define the threshold for any sensitive system or system component. However, the upset of some function or component for exceeding the threshold as a result of EMP induced stimulus applied to that component is not, in general, independent of other stimuli applied simultaneously to other components or subsystems of the same system. That is, there is often an element of mutual coupling or interaction between critical components. This synergistic effect must be taken into account for vulnerability analysis whenever it is appropriate, since the response of one circuit or component, when subjected to EMP, can be significantly different from its response when all circuits or components are excited at the same time.

Multiport excitation can be very important in determining threshold. Some circuits, for example, will not switch if two ports are excited by the same amount simultaneously. Flip-flops have been found which would resist EMP transients of large amplitude if both "set" and "reset" inputs were excited together. In system-level direct drive tests, it is sometimes found that the input wire of some interface circuit is indeed most sensitive. At other times, the most vulnerable element is a sensitive wire deep within the logic-chain which responds to radiated fields. In still other cases, the dominant cause of input is the transient variation of the power supply. Thus, large error in EMP vulnerability analysis, and a complete lack of understanding of the cause of failure, can result if only the inputs to critical components or circuits are assessed. The dominant critical components or chain (circuits) can only be selected after careful study of the behavior of each mission-critical-circuit-chain in its system configuration.

When EMP vulnerability is based on the success or failure of the system mission, the latter will undoubtedly constitute an important factor in vulnerability analysis. Mission critical functions and critical points, failures of which threaten mission failure, vary from system to system. For most critical points and circuits governing critical functions, a burnout caused by EMP leads to system vulnerability. Yet some circuits are harder to burn out than they are to temporarily upset. Thus, it often happens that the upset threshold controls or dominates system vulnerability and, therefore, constitutes the principal basis for vulnerability assessment.

Within a given mission-critical function, there may be many critical-chain circuits. Every input, ground, output, power lead and clock lead of every circuit in that chain is a potential critical point for the upset or failure of that function. If the system were to be assessed in terms of each and every one of these critical points individually, the task of EMP vulnerability assessment might be prohibitive. One, therefore, seeks to identify the most susceptible part of the chain, based on circuit connectivity and sensitivity, and to select these for vulnerability analysis.

5.2 A Conceptual Approach for Susceptibility Estimation

Among the many problems associated with the assessment or evaluation of system vulnerability in a nuclear environment, two basic ones are the determination of the quantitative measures of the causes for the vulnerability and the evaluation of consequential effect of these causes on the degradation of system performance. Causes for EMP vulnerability are electromagnetic fields that penetrate into the system through shields and irregularities of the shields, EMP-induced current and voltages in various cables, circuits, power lines, etc. Quantitative measures of the causes of system degradation can be determined, or at least estimated, from the knowledge of the environment for any given EMP threat, although solutions to such problems are not obvious at times. This is because EMP environment of a system is, in general, not the same as encountered in open space. More specifically, many systems of concern are housed in enclosures that modify the electromagnetic environment and cable currents that would have existed in the absence of such enclosures. Imperfections of the shield also introduce additional changes in the characteristics of the field inside the enclosure. From the knowledge of the field outside the enclosure and various coupling means by which electromagnetic energy can penetrate inside, one can, or at least hope to, estimate the appropriate electromagnetic field characteristics of concern for the system. Furthermore, the amplitudes of EMP-induced voltages and currents at various points in the system can also be estimated from the knowledge of the source and coupling technology as discussed in earlier chapters. Thus, the first problem seems to have a solution in most cases of interest. To determine the vulnerability of a system, one must now establish a correspondence between EMP-induced causes, such as fields, currents and voltages, and their effects which are manifest as system malfunction.

The adverse effect on a component or subsystem from a specified cause may be regarded as a component or subsystem susceptibility which includes temporary or permanent upsets and burnouts. Unfortunately, it is difficult to assign a scientific measure to such an adverse effect or susceptibility. Perhaps one of the basic reasons for this difficulty is due to the ambiguous definition of the adverse effect such as performance degradation or equipment malfunction. Any attempt to determine the susceptibility of a subsystem, therefore, will fail unless a quantitative measure can be assigned to the malfunction or degradation. This is precisely something which cannot be done easily. For example, one can obtain a level of voltage or current that will cause a burnout of a semiconducting device in a receiver. This, however, may not say anything about the receiver vulnerability. For example, if from the viewpoint of mission, the receiver is required to provide an output for 50 microseconds following some $t = o$, one would not miss the receiver even if all the semiconducting devices burn out well after this period. In other words, it is conceivable to have situations in real life where component or subsystem malfunction does not adequately determine system vulnerability, particularly from the mission or operational point of view. In order to denote the performance of a system, it is possible, however, to assign a quantitative measure (number, for example) so that any deviation from this number may represent a malfunction.

To obtain such a quantitative measure of malfunction, it is convenient to start from the end objective. For example, in a receiver, the end objective is to receive a signal at the desired time with an acceptable error in reception. Thus, at a first glance, it may appear that a meaningful and quantitative definition of the malfunction of a receiver may not always be possible. Obviously, when the receiver is only a small part of the system, the overall vulnerability problem becomes considerably more involved.

In spite of the apparent difficulty in obtaining a quantitative measure of malfunctions for EMP vulnerability analysis, one can approach the problem by considering a mathematical model which is amenable to analysis, recognizing that an appropriate judgment has to be made in constructing the model according to the real-life situations.

To illustrate this point, let it be assumed that one is required to determine the effect of EMP on a system which is to (1) detect and identify a target by means of a radar, (2) determine the location of the target, and (3) provide appropriate guidance data to an anti-missile-missile so that it can hit the target with a preassigned CEP (circular-error probability) of hit. Let the quantitative measure of the CEP be C. If there are N, number of input guidance data that affect C, one may write in most cases*:

$$C = \sum_{i=1}^{N_i} a_i \Delta f_i \tag{5.1}$$

where Δf_i denotes the deviation of guidance parameters such as the coordinates and the time-rate of change of the coordinates of the target with the radar at the center of the coordinate system, and a_i is the ith error coefficient which depends on the last stage system characteristics. If it is assumed that each Δf_i is produced by another set of inputs influenced by EMP, one may write:

$$\Delta f_i = \sum_{j=1}^{N_2} b_{ij} \Delta g_{ij} \tag{5.2}$$

where Δg_{ij} is the deviation from such inputs, and b_{ij} is the corresponding error coefficient.

* The linearity assumed in Equation 5.1 is justified for small values of Δf_i.

Similarly, if a number of distinct links in cascade constitute the overall system, one may write:

$$C = \sum_{i=1}^{N_1} a_i \sum_{j=i}^{N_2} b_{ij} \sum_{k=1}^{N_3} c_{ijk} \cdots \Delta p_{ijk} \cdots \qquad (5.3)$$

Equation 5.3 describes the total effect of EMP on the system performance in terms of subsystem parameters, a_i, b_{ij}, c_{ijk}, etc., and EMP-induced perturbations such as Δp_{ijk}, etc. The subsystem parameters can be determined by specific tests. Since,

$$a_i = \left.\frac{\partial C}{\partial f_i}\right|_{f_{io}} \qquad (5.4)$$

that is, the derivative of C with respect to f_i evaluated at f_{io}, which is the normal operating point for the function f_i, one can determine a_i when C as a function of f_i is known, or can be experimentally determined. If all the coefficients are determined and all the input parameters are computed for a given threat, one can obtain a quantitative measure of the system susceptibility due to the threat. The task of obtaining such a measure, however, could be tedious, particularly when there are many links in the system chain. Thus, the method described above may be practical only when a few coefficients and environmental parameters are involved.

It is also possible to formulate the vulnerability problem in terms of statistical probabilities. For example, let a system consisting of a number of essential links as shown in Figure 5.1 be considered. Also, let it be supposed that the operational characteristics of the system are so defined that the overall probability of success p(s) is equal to or greater than a given number N.

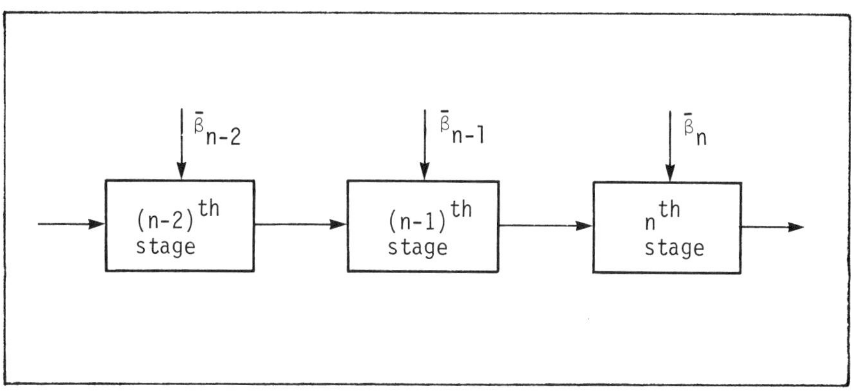

Figure 5.1 - Schematic Representation Of an n-Stage System for Estimating The Probability of Success.

If the output of the final stage that affects this probability is determined by a vector $\overline{\alpha}_n$, one may write:

$$N \leq p(s) = p \left\{ |\overline{\alpha}_n| \leq \varepsilon \right\} \quad (5.5)$$

where n refers to the nth stage, which is the same as the last stage in this case. The vector $\vec{\alpha}_n$, however, can be written as:

$$\vec{\alpha}_n = \vec{F}_1^{\,n}(\vec{\alpha}_{n-1}) + \vec{F}_2^{\,n}(\vec{\beta}_n) \quad (5.6)$$

where $\vec{\alpha}_{n-1}$, denotes the susceptibilities of the subsystem preceding the final stage and $\vec{\beta}_n$ denotes the independent EMP-induced inputs at the last stage (without going through other links or subsystems in the chain) and $\vec{F}_1^{\,n}$, $\vec{F}_2^{\,n}$ are functions of the parameters shown within the parenthesis.

The probability of success p(s) can then be expressed as:

$$p(s) = p \left\{ |\vec{F}_1^{\,n}(\vec{\alpha}_{n-1}) + \vec{F}_2^{\,n}(\vec{\beta}_n)| \leq \varepsilon \right\} \quad (5.7)$$

Again, the vector $\vec{\alpha}_{n-1}$ results from the input of the preceding stage and, in general, can be expressed as:

$$\vec{\alpha}_{n-1} = \vec{F}_1^{\,n-1}(\vec{\alpha}_{n-2}) + \vec{F}_2^{\,n-1}(\vec{\beta}_{n-1}) \quad (5.8)$$

$\vec{\beta}_{n-1}$ being the contribution of the independent EMP-induced inputs at the (n-1)th stage. By a process of mathematical induction, for an n-link chain type system, one has:

$$p(s) = p \left\{ |\vec{F}_1^{\,n} \Big(\vec{F}_1^{\,n-1} \Big(\vec{F}_1^{\,n-2} \cdots \Big(\vec{F}_1^{\,1}(\vec{\alpha}_n) + \vec{F}_2^{\,1}(\vec{\beta}_1)\Big)_1 + \vec{F}_2(\vec{\beta}_2)\Big)_{n-1} \cdots \vec{F}_2^{\,n-1}(\vec{\beta}_{n-1})\Big)_n + \vec{F}_2^{\,n}(\vec{\beta}_n)| \leq \varepsilon \right\} \quad (5.9)$$

In order to make a system invulnerable to a specified EMP threat, it is necessary to make p(s) equal to or greater than N as indicated in Equation 5.5. This, in turn, is achieved by satisfying the criterion:

$$\left\{ \left| \vec{F}_1^{\,n} \left(\vec{F}_1^{\,n-1} \left(\vec{F}_1^{\,n-2} \cdots \left(\vec{F}_1^{\,1} (\vec{\alpha}_o) + \vec{F}_2^{\,1} (\vec{\beta}_1) \right)_1 \right.\right.\right.\right.$$
$$\left.\left.\left.\left. + \vec{F}_2^{\,2} (\vec{\beta}_2) \right)_2 \cdots \vec{F}_2^{\,n-1} (\vec{\beta}_{n-1}) \right)_{n-1} \right)_n + \vec{F}_2^{\,n-1} (\vec{\beta}_n) \right| \leq \varepsilon \right\} \tag{5.10}$$

Intuitively, one may expect such a result as given in Equation 5.10 since the total error in a linear system should be made of various individual stage-errors multiplied by appropriate weighting functions. These weighting functions will depend on the design of individual equipment or subsystem constituting various links on the system.

When the input vectors $\vec{\alpha}_1$, $\vec{\beta}_1$, etc., are known, Equation 5.9 will denote a definite, instead of a probabilistic, measure of success. In other words, if the total error resulting from EMP-induced inputs is less than the tolerable limit ε, the probability of success, insofar as EMP effects are concerned, is one. However, in practice, one may have to express the inputs $\vec{\alpha}_1$, $\vec{\beta}_1$, etc., in statistical forms with known or assumed distributions. EMP vulnerability of the system, then, can be determined only probabilistically.

The theoretical approaches for vulnerability analysis as outlined above are suitable for small but sensitive systems where the error coefficients a_i, b_{ij}, etc. or the functionals such as $\vec{F}_1^{\,n}$, $\vec{F}_2^{\,n}$, etc., are known or can be determined readily. For a complex system involving many stages and "feedback loop", it is desirable to formulate a simplified vulnerability analysis approach based on the knowledge of the system, its mission and EMP threat, along with appropriate engineering judgments, and a series of relatively simple analytical tools on coupling, system sensitivity, etc. Such an approach is outlined in the following section.

5.3 Vulnerability Analysis Methodology

For most electronic weapon systems, it is possible to follow a step-by-step methodology to assess or predict EMP vulnerability and to formulate remedial means which may harden a given system against a specified EMP threat or threats. Essential steps for this purpose are:

- Information Assembly.
- Physical Inspection and Qualitative Evaluation.
- Zoning.
- Coupling Analysis.
- System Performance Analysis and Vulnerability Prediction.

5.3.1 Information Assembly

From a pragmatic viewpoint, EMP vulnerability of a system should depend on the physical characteristics of the system, system mission, and characteristics of EMP threat or threats. In the information assembly process, then, one should assemble all pertinent information on these three aspects. Typical information on the physical characteristics of the system may be obtained from:

a) System and subsystem block diagram,
b) Auxiliary system (power, cooling, test, grounding, communication), their schematics and wiring diagram,
c) Functional system description,
d) Selected schematic diagrams of sensitive components and subsystems,
e) Subsystem wiring diagram,
f) Site plot plan,
g) Floor layouts (when applicable) with equipment locations shown,
h) Blueprints of building housing the system having pertinent construction details,
i) Design criteria manual (EMC, EMI, RFI specifications).

Since vulnerability, by definition, is directly related to the system mission, one must define this mission at the outset. Many systems have a singular mission which can be defined readily. Some systems, however, may have multiple missions. EMP vulnerability analysis for such cases have to be carried out for each of the missions separately.

Also needed in the information assembly process is the definition of EMP threat. Although the actual EMP environment, as noted in the previous section, depends on many factors, only the worst-case threat need be considered for the vulnerability analysis, since one may not depend on any less stringent environment. Besides, a system hardened for the worst-case threat will obviously be hardened for the entire threat spectrum.

In addition, one should obtain specifications of necessary protection level, sensitivity of components, nature of interconnectivity of subsystems and auxiliary systems, etc., during the information assembly.

5.3.2 Qualitative Evaluation

The next important step in the vulnerability assessment methodology is the qualitative evaluation of the system. For a system in existence, this involves a physical examination to assure satisfactory compliance with good design rules and practices, location of weak points, particularly those which may affect the mission, etc. This step is important since it is pointless to determine the nature of adverse effect on a system due to an obvious violation of good design rules and practices, when such a violation should be removed or repaired as a matter of course, and when it is more expensive to analyze the system with a violation than to repair it. Also, many weak points and violations are not evidenced from the block diagram, schematics and construction drawings.

During the physical inspection, the observed violations of good design rules may be classified as:

- Acceptable in this particular instance.
- Questionable and require further examination or testing.
- Questionable but easy (not expensive) to repair (this class of violation is repaired as a matter of course).
- Clearly not acceptable and must be repaired.

Typical examinations which are included during the physical evaluation of a ground-based system, for example, are:

- Examination of the outside zone.
- Grounding system.
- Examination of the major structure.
- Examination of cabling and wiring.
- Interior grounds.
- Equipment layout and location.

For outside zone examination, typical tasks are:

a) Procurement of a plot plan of the ground surrounding the facility,
b) Marking in one color the pertinent EM structures, pipes, cables, conduits and buildings,
c) Physical examination of the junction points which might complete loops,
d) Marking the complete loops by a second color,
e) Location and flagging of other lines, pipes, cables, conduits, etc., which might be exposed to strong EMP fields,

f) Marking these lines by a third color,

Typical tasks in grounding system examination are:

a) Sketching the grounding system by using one of the architecural drawings and plot plans,
b) Indication with a color code where various levels of system connect to ground systems. All phantom grounds, shield grounds, and earth grounds are to be included for this purpose.
c) Marking the bonding points,
d) Marking the common signal paths,
e) Marking the ground loops,
f) Indication of ground impedances,
g) Flagging the violations,

Similarly, the tasks for examining the major structure may include:

a) Preparation of several transparent copies of both cross section and plan view of architectural drawings and marking the boundaries of the shielded areas.
b) Obtaining auxiliary system drawings such as those of power, sewage, cooling, water, etc.
c) Inspection of the system to find shielding violations such as seams, cracks, doors, vents, welds, common walls, materials used and design defects.
d) Location of EMP-effect penetrations from auxiliary system drawings and marking them onto architectural drawings.
e) Inspection of the facility and classification of penetration violations that actually exist.

Tasks for the examination of cabling and wiring may include:

a) Classification of the cabling by level and functions, and allocation of an identification color to each class.
b) Mapping and marking of cable trays and cabling runs on a copy of the architectural drawing.
c) Marking the drawing by color to show the construction and type of cable used.
d) Flagging the cable shielding and grounding violations.
e) Flagging the coupling violations.

For interior grounds one may examine interior equipment grounding and landing in the same manner as exterior grounds. The task breakdowns for the examination of equipment layout and location may be as follows:

a) Sketching the equipment racks and labeling them by function on the architectural drawings.

b) Classification of the equipment by sensitivity, allocating different colors for such classes.
c) Marking the circuit and equipment layout violations.

The use of colors as suggested above is helpful for identifying different classes of problems that need to be analyzed for the vulnerability assessment. Of course, not all tasks referred to above are applicable or meaningful, even for all ground-based electronic or weapon systems. The suggested tasks, however, demonstrate the required depth of the physical examination.

5.3.3 Zoning

The next step in EMP vulnerability analysis is zoning. The purpose of zoning is to facilitate such an analysis by:

a) Identifying those volumes of space for which an estimate of EMP-induced electromagnetic environment must be obtained in order to determine the effect of a worst-case nuclear detonation on system performance.

b) Defining the specific coupling problems that must be solved in order to obtain an estimate of EMP-induced electromagnetic field environment in the volumes of space identified in (a).

c) Locating and grouping equipments, in a new system design, to minimize the number and size of shielded enclosures required, and to locate sensitive equipments in regions of low field strength.

The types of zoning to be performed in conducting vulnerability analysis of an existing system are:

1) Sensitive area zoning.
2) Environmental zoning.

Sensitive area zoning involves the location and demarkation of areas containing functionally sensitive equipment. A functionally sensitive equipment is one which is vital to proper system operation and which may be susceptible to upset by EMP-induced electromagnetic environment. The purpose of sensitive area zoning is to designate those areas for which an estimate must be obtained of EMP-induced electromagnetic environment. A description of the physical location, and the boundaries of sensitive areas and the interface connections, such as cables, is required for this purpose.

Sensitive area zoning is conducted on a subsystem basis. The sensitive areas of each subsystem are located and marked. The cables which connect each sensitive area with the rest of the system are also marked. Because the zoning is done on a subsystem basis, the sensitive areas of different subsystems may overlap or even have the same boundaries.

Environmental zoning involves the definition of electromagnetically distinct regions. In general, these are volumes of space at whose boundaries there is an abrupt change in the electromagnetic field environment. For example, the interior of a shielded building or a shielded room is an electromagnetically distinct region. The purpose of environmental zoning is to define those regions or volumes of space, for which separate coupling calculations must be performed. Each functionally sensitive area will be inside one of the environmental zones. An environmental zone may enclose a number of functionally sensitive areas.

Environmental zoning may be primary or secondary. A primary environmental zone is one at whose external boundary the incident electromagnetic field is the ambient EMP field. Inside the boundary, the field will be significantly reduced in amplitude. A secondary environmental zone is one completely enclosed by a primary zone. For example, a shielded room inside a shielding structure defines a secondary environmental zone. To calculate electromagnetic environment inside a secondary zone, it is necessary first to calculate the field inside the primary zone. This modified field is then used for further calculations to determine the field inside the secondary zone.

For a new system, the zoning, which consists of grouping equipments in such a manner as to minimize EMP coupling to sensitive subsystems, and to minimize the required number and size of shielded enclosures for an overall invulnerable system, itself becomes a part of the design process. The equipment grouping in such a case may be effected within various iso-susceptibility boundaries such that one level of protection is furnished to all equipment inside one boundary. EMP protection can then be afforded by enclosing each iso-susceptibility region in a separate shielding enclosure, with an appropriate treatment of interconnecting cables, or by wiring a single enclosure with adequate wall attenuation to enclose all regions, or by providing one or more small shielding enclosures within a large enclosure to protect particularly sensitive iso-susceptibility regions.

As in the case of qualitative (physical) examination of a system, one can formulate a set of tasks for the zoning procedure. They are, for example:

a) Sketching in boundaries that enclose areas containing functionally sensitive equipment on the equipment layout drawings referred to in "Qualitative Evaluation". This will be feasible since, during the qualitative evaluation, the equipment racks will already be labeled by function and classified by sensitivity. Different colors may be used for each subsystem, and only one subsystem should be used at a time. While delineating the functionally sensitive areas, sensitive areas of each subsystem should be numbered, although sensitive areas of different subsystems may overlap.

b) Identifying connections and penetrations of each functionally sensitive area, including cables and other structures (such as pipes), which may conduct EMP-induced electric currents into the area.

c) Sketching on the equipment layout drawings, the path of cables which penetrate into each functionally sensitive area. Each cable should be numbered and labeled as to its function and signal level.

d) Outlining, numbering and labeling environmental zones on the plan view and cross section architectural drawing, referred to in "Qualitative Evaluation".

e) Preparing one chart which designates the environmental zone in which each of the functionally sensitive areas is located (using the drawings showing functionally sensitive areas), and another chart designating the environmental zones traversed or penetrated by each of the cables identified and numbered in (c).

Again, all such tasks will not be appropriate for all systems, but they represent the nature and depth of the tasks to be performed for zoning.

5.3.4 Coupling Analysis

One of the essential steps in vulnerability analysis leading to the determination of EMP-induced currents, voltages and fields is the coupling analysis. A complete description of the "ambient" EMP environment produced by the worst-case threat is the starting point for such an analysis. This description includes the time and amplitude characteristics of EMP fields above and below the surface of the earth throughout the region of space occupied by the system under consideration. The consideration of the problems of shielding, aperture coupling, antenna currents and voltages, and cable currents constitutes the basic tasks for the coupling analysis.

As discussed in Chapter 3, the analytical problems involving transient electromagnetic coupling technology can be solved only in certain idealized cases. The complexity of the EMP coupling problem for shielding structures of arbitrary geometry with various apertures, imperfections, etc., suggest that such problems be resolved through appropriate judgment based on the results of analytically solvable problems and similarities between the idealized structure and the structure of concern.

Principal tasks in coupling analysis are:

a) Determination of EMP-induced fields in each primary environmental zone due to penetration through shielding walls.
b) Determination of EMP-induced fields in each primary environmental zone due to apertures.

c) Determination of EMP-induced fields in each secondary environmental zone.
d) Estimation of currents in cables which connect to sensitive equipments.
e) Determination of antenna currents and voltages.

As subtasks under (a), one should:

i) Equate each primary and secondary environmental zone which is defined by a solid shielding wall to an approximately similar idealized geometry (sphere, long cylinder, parallel-plate structure, etc.). For example, a cubical structure more closely resembles a sphere than a long cylinder. A long structure, whose cross section is approximately a square, resembles a long cylinder, etc.

ii) Evaluate, for solid-wall enclosures the constants B and C in MKS units where:

$$B = d\sqrt{\mu\sigma}$$

$$C = 2\frac{\mu d}{\mu_o b}$$

for each environmental zone, d, μ, σ, μ_o and b being wall thickness, permeability of wall material, conductivity of wall material, permeability of freespace and radius of equivalent cylinder respectively. For a sphere, or nearly spherical geometry, b is 2/3 of the sphere radius. For a parallel plate-like structure, b is the same as the separation distance.

iii) Use parametric graphs given in Figure 3.13 to obtain field inside each solid wall enclosure for a unit delta function applied field.

iv) Determine the time-varying field inside the enclosure by using convolution integral of the data-function response in (iii) and the actual time-varying incident field. Thus, if the actual time varying incident field is given as g(t), the field inside the enclosure will be:

$$\int_{-\infty}^{\infty} f(t-T)\, g(T)\, dT$$

f(t) being the delta function response as referred to in (iii).

To determine EMP-induced field in each primary environmental zone due to apertures, it should be noted first that all apertures of concern in each environmental zone would have been flagged and examined during the qualitative evaluation of the system. In the event that any of the apertures have not been properly handled or have not been fixed, calculations have to be performed to evaluate their effects. Figures 3.15 and 3.16 may be used to estimate the field penetration due to apertures. Finally, the aperture field is to be added to the field due to wall penetration to obtain the total field inside the environmental zone. If aperture effects contribute significantly to the internal field, the iso-environmental contours must be changed accordingly.

The same procedure is followed to determine the fields and iso-environmental contours inside the secondary zones. For these zones, however, the ambient or incident field is the field inside the environmental zone which surrounds the secondary zone.

The problem of estimating currents in cables which connect to sensitive equipment can be divided into three domains. The first of these includes cables which are completely enclosed inside an effective shielding structure which reduces the ambient EMP field by a number of orders of magnitude. The second includes buried cables which, over at least part of their lengths, are subject to the subsurface ambient EMP environment. The third includes cables which are subject to the above-ground ambient EMP environment. These classifications, however, are not unique and are not applicable to non-ground based systems.

In the case of well shielded cables inside building structures, it may not be necessary to carry out complex calculations to estimate the current induced on the inner conductors. For example, if the shielding effectiveness of the structure is known, a rough estimate can be made on the induced current by conventional techniques. If these pulses are negligible in comparison with the amplitudes of signals carried by the cables of concern, further calculations are not necessary. Even for such cases, care must be exercised while considering EMP-induced currents in coaxial cables which connect two pieces of equipment and frequently form a large ground loop. The transient magnetic field of EMP can cause a large circulating current in the outer conductors of such cables.

Extremely large currents can be generated in the sheaths of both buried and above-surface cables which are subject to the ambient EMP environment. These sheath (or outer conductor) currents, in turn, induce currents in the inner conductors which may terminate to functionally sensitive equipments. If proper design practices are carried out, the large currents in cable shields will not be conducted into the shielded primary or secondary environmental zones. However, if the cable shields do penetrate into environmental zones without proper treatment, it will be necessary to obtain an estimate of the characteristics of the current pulse. In any event, it is necessary to obtain an estimate of the current carried on the inner conductor due to EMP field.

A discussion of the problems of induced-cable currents due to EMP is presented in Section 3.6. Unfortunately, there is no single source of engineering solutions to the cable current problems which are encountered in performing EMP vulnerability analysis. Various solutions discussed in Section 3.6 and appropriate engineering judgment, however, will constitute a convenient starting point for this purpose.

A principal subtask toward the estimation of cable currents is to ascertain whether each cable of concern is a:

- Long exposed or underground cable where EMP environment varies over the length of the cable.
- Short exposed cable, where EMP environment is constant over the length of the cable.
- Long insulated cables.
- Short insulated cables.

To obtain cable currents, expressions for appropriate cable parameters such as characteristic impedance, length, conductor dimensions, series impedance, etc., should be used as needed.

An estimation of the currents and voltages induced in antennas is readily obtained if the antenna transfer function $T(\omega)$, and its load transfer function $F(\omega)$ are known, since the antenna response in such a case is the inverse transform of $F(\omega)G(\omega)T(\omega)$, $G(\omega)$ being the Fourier transform of EMP incident field. Discussions on antenna voltages and currents due to EMP field are presented in Section 3.2. Since $T(\omega)$ and $F(\omega)$ are different for different types of antennas, the first subtask for estimation of antenna currents or voltages is to ascertain the type of antenna, such as a dipole in free space, dipole over a ground plane, monopole, loop, linear array, etc., if the antenna is linear, and horn, dishes, etc., for aperture antennas. Results presented in Section 3.2 for the specific type of antenna of concern should then be used to determine EMP coupling through antennas as needed for the vulnerability analysis.

The last step in EMP vulnerability analysis is system performance analysis and vulnerability prediction. Susceptibilities in electronic system and subsystems such as latch-up, burnout, etc., and circuit analysis which could be performed for determining system performance and vulnerability prediction, are discussed in Chapter 4. Essential tasks involved for this purpose are the selection of sensitive subsystems or components for which the analysis must be performed, and executing such analysis with various tools presented in Chapter 4.

EMP vulnerability analysis is often a tedious task and there is seldom a shortcut to lessen the amount of work involved, particularly when a high degree of system invulnerability or hardness is desired. An extreme thoroughness in examining every subsystem, which could be sensitive either directly due to EMP or due to synergistic effects of sister subystems affected by EMP is required. Although a wide range of uncertainties may exist for EMP threat and ascertaining system parameters accurately, and the analytical tools are often limited, it is believed that an acceptable EMP vulnerability prediction is feasible for most systems, if not all.

5.4 References

1. Brown, G. L., Ghose, R. N., Green, J. X., and Wouters, L. F., *Electromagnetic Pulse Energy Coupling Studies*, Vol. III, American Nucleonics Corporation, Glendale, California, September 1967.

2. *Electromagnetic Pulse Handbook for Missiles and Aircraft in Flight EMP Interaction 1-1*, Air Force Weapons Laboratory, September 1972.

3. Tesche, F. M. and Castillo, J. P., *Computational Techniques for EMP Interaction*, Third Symposium and Technical Exhibition on Electromagnetic Compatibility, Rotterdam, May 1979.

CHAPTER 6

Hardness Concepts and Techniques

EMP susceptibility in electronic and weapons systems can be substantially reduced by effectively reducing the EMP coupling into such systems. The reduction of coupling can be effected by shielding sensitive components and subsystems with a metallic enclosure wherever practical. Obviously, the entire system cannot be shielded since antennas or other collectors of signals which must interface with the system may not perform their intended functions when completely enclosed within a metallic shield. Other means of reducing EMP coupling include filtering which rejects EMP energy at frequencies outside the frequency band of interest for the system. EMP coupling reduction can also be effected by voltage limiting, good design practices and other EMP circumvention techniques which are further discussed in this Chapter.

6.1 Shielding

In the previous chapters, the sources and mechanisms of the nuclear EMP, its coupling into the electronic or weapon systems, and various effects of such couplings have been considered. When EMP is a threat, its source and mechanism of generation is often outside the control of the victim system. Various means and methods, however, are known to reduce the coupling of EMP into the victim system and hence to minimize the consequential susceptibility. An obvious method of denying the penetration of EMP energy into a system is to provide shielding by a continuous metallic enclosure.

In an idealized shielding enclosure, the reduction of EMP field strength inside the enclosure is determined primarily by the reflection loss at the outer surface where the EMP field impinges on the enclosure, and the exponential attenuation of the field that occurs during its penetration through the shielding material. Since this attenuation is 8.68t dB where t is the thickness of the shielding material in terms of its skin-depth, a continuous thick metallic enclosure can reduce the EMP field inside the enclosure to any arbitrary level, when appropriate thickness of the shielding material is provided. Shielding, therefore, could be an effective tool to minimize EMP energy coupling into a system and hence to minimize its EMP vulnerability.

Most classical considerations of shielding involve the determination of electromagnetic field inside simple geometric structures, such as a spherical shell, or a finite closed cylinder, or a rectangular box, when an arbitrary electromagnetic field impinges at the outer surface of the structure. Associated analyses for such cases involve the solution of the boundary-value problems at the outer and inner surfaces of the structures for the specified impinging field due to the EMP.

6.1.1 Physical Mechanism of Shielding

The physical mechanism of the field penetration through a shielding structure can, perhaps, be explained by considering an idealized enclosure having no openings and constructed of a continuous sheet of conducting, and possibly permeable, material. For the quasi-static electric field impinging at the outer surface of the enclosure, no electric field appears within the enclosure. When, however, the impinging electric field varies with time, the electric charges redistribute themselves at the outer surface of the enclosure and a current flows at the outer surface. Since the charge distribution is proportional to the applied electric field, the current will be proportional to the derivative of the charge of applied electric field. As a result, the magnitude of the current flowing on the surface of the enclosure, and the energy appearing inside the enclosure, will be directly proportional to the frequency of the applied electric field. Thus, at very low frequencies, the currents induced into the structure by time-varying electric fields are very small. At sharp corners of the enclosure, there is a concentration of charge, and this in turn causes the current flow to be concentrated near the corners.

If the impinging field is quasi-magnetic instead of quasi-electric as considered above, and the enclosure is made of a highly permeable material, some of the applied magnetic field will be ducted by the enclosure walls, provided the enclosure material is not saturated. Usually, a good magnetic shielding is obtained only by a very thick-walled enclosure. For the low frequency time-varying magnetic field impinging on a thin-walled enclosure constructed of a highly conducting material such as aluminum, there is a current flow, although most of the current flow is near the edges of the enclosure. Thus, the enclosure structure, in effect, becomes a "shorted turn" of a transformer. The magnetic field in this case, induces a voltage integral around the shorted turn which is proportional to the derivative of the magnetic field. At very low frequencies, where the inductive reactance is negligible, the current flowing in the loop is proportional to that voltage divided by the resistance of the loop. This current, however, is relatively small and is in phase with the applied magnetic field. As a result, little cancellation or reflection of the applied magnetic field occurs. As the frequency of the applied magnetic field increases, the inductive resistance dominates over the loop resistance, and the current flow in the loop increases. This current is 90-degree phase-shifted with respect to the applied magnetic field, and the reactive magnetic field due to the current is 180-degree out of phase with respect to the applied magnetic field. Some cancellation and reflection of the incident magnetic field thus occurs inside the shorted turn.

As the energy of the incident field is increased, a significant amount of this energy is absorbed by the walls of the enclosure by the skin-effect mechanism. This causes the time-varying electric or magnetic field incident on the outside of the enclosure to be attenuated exponentially before reaching the interior surface. The attenuation of induced currents is given by:

$$i \simeq i_o e^{-d/2s} \tag{6.1}$$

where*, $s \simeq \dfrac{1}{\sqrt{\mu_o \sigma \beta}}$ for a pulse

$\dfrac{2}{\sqrt{\mu \mu_o \sigma \omega}}$ for a discrete frequency

---

* For the discrete frequency, s is the real part of the propagation constant for the enclosure wall material, that is, Re $(\omega^2 \mu\, \mu_o \varepsilon\ - i\, \omega \mu \mu_o \sigma\ )^{\frac{1}{2}}$, ε being the dielectric constant of the material.

d	=	the wall thickness
ω	=	2 x frequency of the incident field
μ_o	=	permeability of the free space
μ	=	magnetic permeability of the wall material
β	=	pulse width for a Gaussian pulse ($e^{-\beta^2 t^2}$)

6.1.2 Estimation of Shielding Effectiveness

The shielding effectiveness or the figure of merit of a shield is a measure of the degree of attenuation obtainable due to a shield. By definition, the shielding effectiveness, S.E., is:

$$S.E._{dB} = 10 \log_{10} \frac{\text{incident power density}}{\text{transmitted power density}}$$

White* has presented a comprehensive discussion on shielding concepts, approaches, and design parameters for shields useful for many purposes including EMP hardness. As noted in this handbook, the attenuation provided by a shield consists of three parts. The first part involves the reflection loss resulting from the mismatch of the wave impedance and the intrinsic impedance of the shield material. Following the reflection at the shield surface as the wave impinges on it, a part of the incident wave energy is transmitted into the shield. As this transmitted field penetrates through the shield, usually the non-zero conductivity of the shield material causes the field to attenuate exponentially with the distance traversed by the field inside the shield material. This attenuation constitutes the second part of the S.E. loss, and is referred to as the absorption loss. The third part of the S.E. loss, termed the re-reflection correction, is encountered due to the mismatch between the metal impedance and the wave impedance at the other side of the shield following the penetration. White has shown that for a shield having a thickness d,

$$\text{the absorption loss} = A_{dB} = 20 \log_{10} e^{\alpha d} = 8.686\ \alpha d$$

$$\text{the reflection loss, } R_{dB} = 20 \log\ (1 + k)^2/4k$$

and the re-reflection correction,

$$B_{dB} = 20 \log_{10} \left| 1 - \left[(k - 1) / (k + 1)\right]^2 e^{-2\gamma d} \right|$$

where, $k = Z_w/Z_m$

* D. R. J. White, "A Handbook on Electromagnetic Shielding Materials and Performance", Don White Consultants, Inc., 1980.

$$Z_w = \text{wave impedance} = \sqrt{\mu_1/\varepsilon_1}$$

$$Z_m = \text{metal impedance} = \sqrt{i\omega\mu_2/\sigma_2}$$

$$\gamma = \left[\omega^2 \mu_2 \varepsilon_2 - i\omega\mu_2\sigma_2\right]^{1/2} = \alpha + i\beta$$

μ_1, ε_1, are the permeability and dielectric constant of the material on either side of the shielding, μ_2, ε_2 and σ_2 are respectively the permeability, dielectric constant and conductivity of the shielding material, and ω is 2π x frequency of the impinging wave.

For most shielding problems, the reflection and re-reflection correction become fixed as the shielding material is chosen. The absorption loss, however, can be increased indefinitely by increasing the shielding thickness.

To simplify shielding effectiveness computations, the effect of the electric field in the lower-frequency region may generally be ignored. It turns out that for most square or box-like enclosures, energy penetration is essentially via the magnetic field. This is not the case for a long structure which is parallel to the electric field.

Another factor in establishing effectiveness is the geometry of the structures. In general, for equal wall thickness, the larger the structure the greater the shielding effectiveness at the center of the structure. Obviously, this will not be true at points closer to the exterior of the enclosure. Certain smooth geometries, such as a sphere, exhibit a uniform field in the interior of the enclosure. Cornered structures on the other hand exhibit poor shielding near the corners, due to the concentration of surface currents at the edges.

The method of determining shielding effectiveness for enclosures of idealized geometrics from the solution of boundary value problems has been discussed in Chapter 3. Such solutions are helpful for the design of shielding enclosure parameters, such as dimensions, geometry, materials, etc., required for a specified shielding effectiveness. It has to be noted, however, that real-life enclosures, such as cabinets for electronic equipment which needs to be protected from EMP, depart widely from the idealized enclosures. In general, these departures severely degrade the shielding effectiveness of the enclosures. Slots and slits or imperfect bonds at enclosure walls also severely reduce the shielding effectiveness.

Furthermore, in many enclosures the essential purpose of providing shielding is defeated by EMP energy carried by cables or pipes penetrating the enclosure.

Good shielding efficiency against electric fields is readily obtained by the use of high-conducting metals such as copper and aluminum for enclosure walls. The shielding against magnetic fields, particularly at low frequencies, however, becomes more difficult since the reflection loss at the boundary, that is enclosure walls, may approach zero for some combinations of materials and frequency. Also, with decreasing frequency, the absorption and reflection losses for non-magnetic materials decrease steadily, thus reducing the shielding effectiveness. Such problems, however, do not exist at high frequencies where the reflection loss at the wall surface boundaries becomes appreciable again, and the skin effect provides a rapid attenuation of the penetrating field. Typical penetration and reflection loss for various metals at different frequencies are shown in Figure 6.1a. Absorption losses for different thicknesses of iron and copper shieldings at different frequencies, suitable for shielding design as provided by White, are shown in Figures 6.1b and 6.1c respectively.

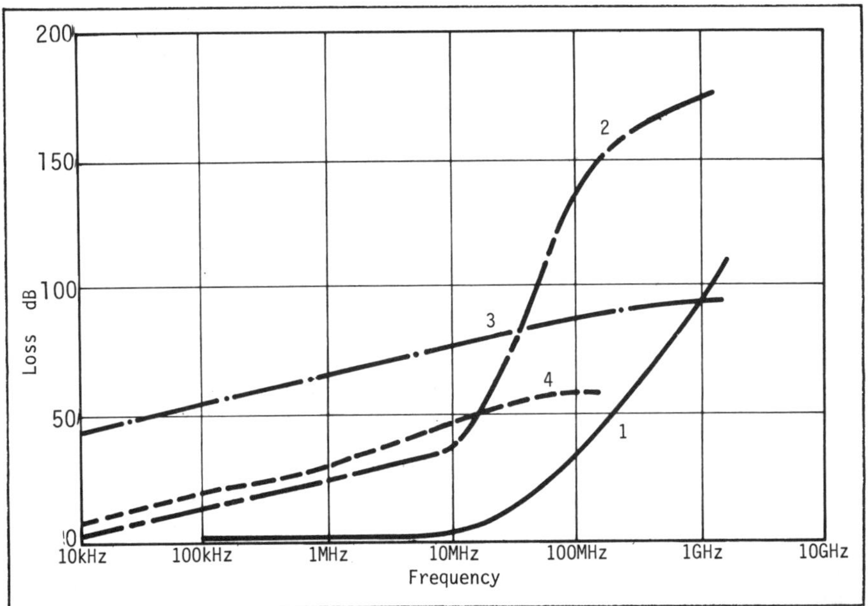

Figure 6.1a - Typical Penetration and Reflection Losses for Various Metals at Different Frequencies.

1. Penetration loss per mil thickness - copper
2. Penetration loss per mil thickness - iron
3. Reflection loss, magnetic field - copper
4. Reflection loss, magnetic field - iron

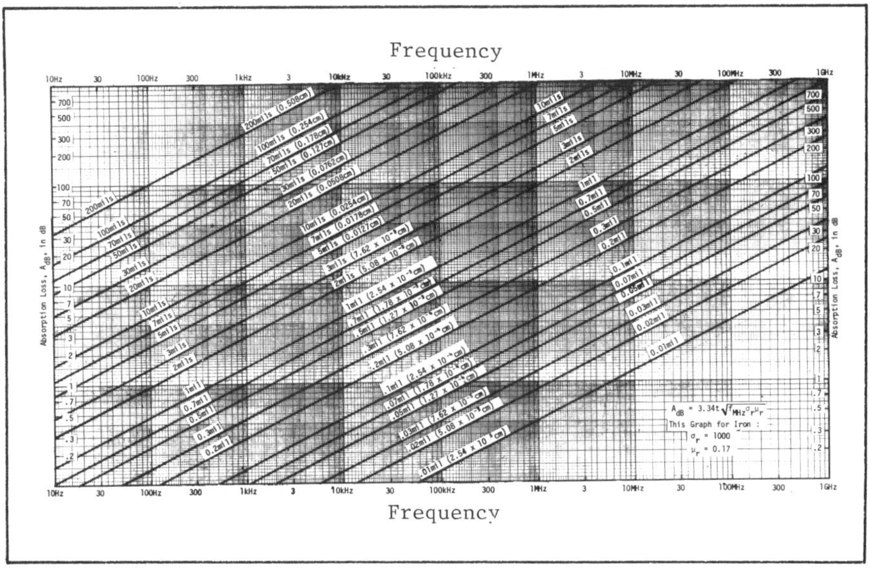

Figure 6.1b - Absorption Loss For Various Frequencies and Thicknesses Of Iron.

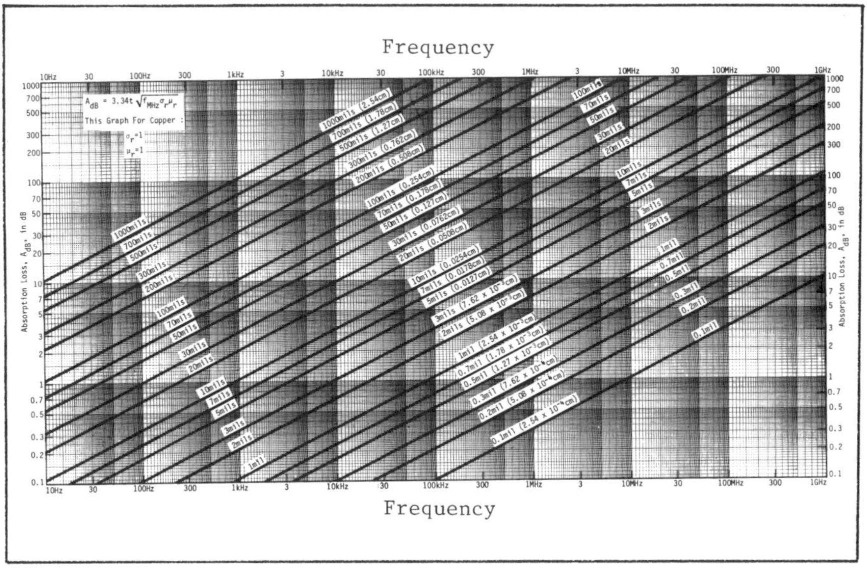

Figure 6.1c - Absorption Loss For Various Frequencies and Thicknesses Of Copper.

6.1.3 Shielding Approaches and Various Shielding Constructions

To minimize EMP coupling in electronic or weapon systems, different approaches for shieldings can be used. The first approach involves the shielding of the building or selected areas of the building housing the equipment of concern. This type of shielding is effective when individual cabinets or subsystems cannot be shielded because of practical considerations. In the second type of shielding, discrete shielding enclosures are used for a separate sensitive element or group of elements in a system. In addition, individual equipment racks, their interconnecting cables, and even compartments within the cabinets, can be separately shielded.

For shielding a building or specified areas of a building, solid sheet metal such as copper or bronze, assembled around a wood framework, can be used. For most purposes, however, galvanized sheet steel has been found to have better properties than copper and is considerably less expensive. When such steel is zinc coated, the high-frequency attenuation characteristics and corrosion resistance are improved. Another desirable characteristic of zinc coating is that under moderate amounts of pressure, it makes good contact closure at panel or door joints.

The shielding effectiveness of a large enclosure using solid sheet metal, as discussed above, depends to a large extent on the care with which such enclosures are constructed, regardless of the type of metal used for the purpose. For example, bad construction practices, such as the use of non-plated or cadmium-plated nuts to mount filters, mounting both through two walls of the enclosure, making unintentional grounds, burning holes on walls, etc., could seriously degrade the effectiveness of an otherwise good shielding enclosure. Cracks and holes need to be avoided in shielding construction, and components within shielded enclosures should be as centrally located as possible, certainly away from corners where current concentrations may exist*.

To design a building-like enclosure with a shielding effectiveness of about 100 dB in the frequency range of 60 Hz to 1 GHz or above, one should use magnetic steel having a relative permeability on the order of 1000 and a thickness of at least 300 mils. All access doors for the enclosure should have finger stock installed around the periphery of the door to make good electrical contact between the contact fingers and door jambs. Filters will be required for all incoming power and instrumentation leads. All penetrations through the floor, walls, and ceiling should be welded around the penetration area. Air ducts, if any, should be so designed that there would be no leakage through the duct.

---

* An illustration of means necessary to increase shielding effectiveness of an equipment case, for example, is shown in Figure 6.6 .

For the discrete shielding enclosure suitable for shielding or protecting equipment or elements of a system one could use commercially available enclosures. One class among them is the so-called Zero ambient class which provides a high degree of attenuation, on the order of 100 dB, for the frequency range of almost dc to 10 GHz. This class of enclosure is made of 1/4- to 1/2-inch-thick iron with welded joints. Although Zero heavy duty cabinets are primarily designed for EMI shielding, they could be very effective for EMP shielding. Typical shielding attenuation of electric and magnetic fields provided by such steel-shielded cabinets over a wide frequency band of interest are shown in Figures 6.2 and 6.3. It is seen that the electric field attenuation is more or less uniform over very wide frequency bands, whereas the attenuation of the magnetic field is not very high at very low frequencies although the attenuation increases with increasing frequencies.

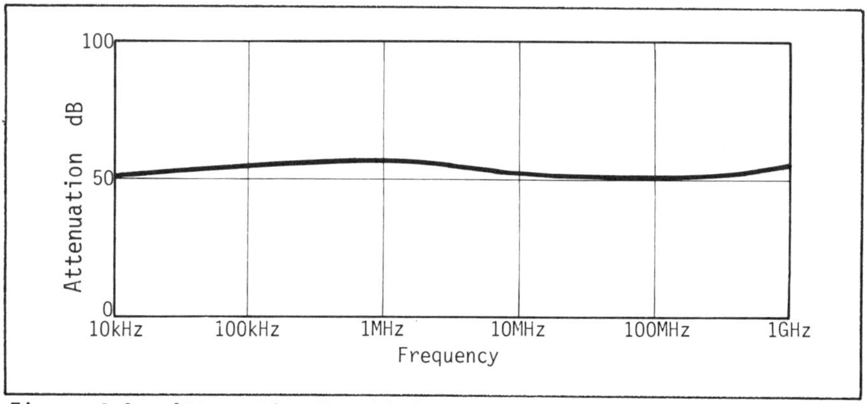

Figure 6.2 - Attenuation Provided by Steel Shielded Cabinet - Electric Field (E).

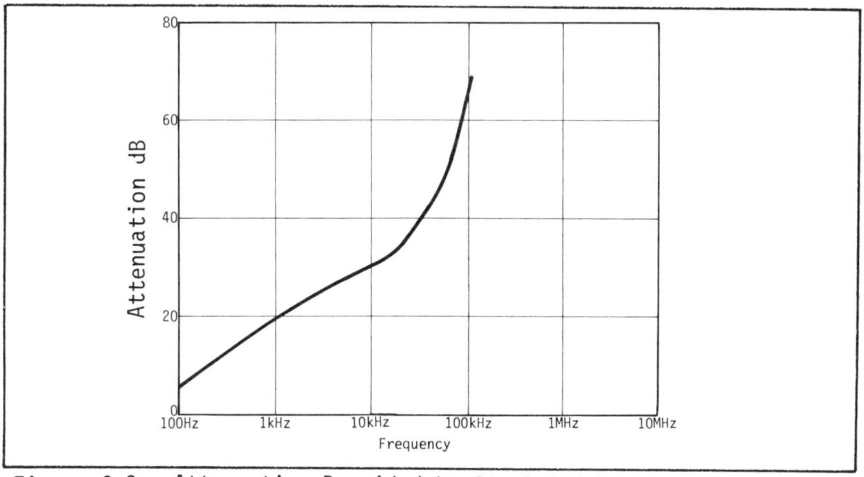

Figure 6.3 - Attenuation Provided by Steel Shielded Cabinet - Magnetic Field (H).

Commercial shielding enclosures are also available in modular forms which can be assembled at the site. They consist of single or double panels with various kinds of joint designs, door designs and corner treatments to maintain shielding effectiveness. The degree of attenuation to electric and magnetic fields vary for such enclosures. However, they can be used effectively for EMP shielding when appropriate enclosures are selected on the basis of the shielding requirements.

As stated earlier, the essential purpose of providing shielding may be defeated when EMP energy is carried into the enclosure by cables or pipes penetrating into such an enclosure. This is because EMP can reach an electronic or weapon system not only by means of direct penetration of the enclosure but also by the induction of currents in cables, pipes, and lines that connect the system to the outside world. It is often necessary, therefore, to provide shielding to such system-inlets to minimize EMP coupling into the system. Isolated copper conductors have very low attenuation to signals propagated along the conductors, particularly those with good insulation. Power lines are especially susceptible to this problem. Unattenuated pulses have been known to damage heavy equipment such as a distribution transformer after propagating several miles. A possible answer to this is the use of a special lossy line which has a relatively low-loss at power line frequencies.

For an effective shielding of a cable, the outer shield should be continuous with a complete enclosure. Copper braided shielding of the type commonly used to shield high-impedance circuits is very nearly useless for EMP protection. The combination braids, made partly of magnetic materials, are better and provide, under some conditions, attenuations of 30 to 50 dB. Where 80 dB or more attenuation is required, continuous solid shields or a combination solid/braid must be used. Shielding effectiveness of solid copper and aluminum sheaths for various 50 ohm coaxial cables is shown in Figures 6.4a and b.

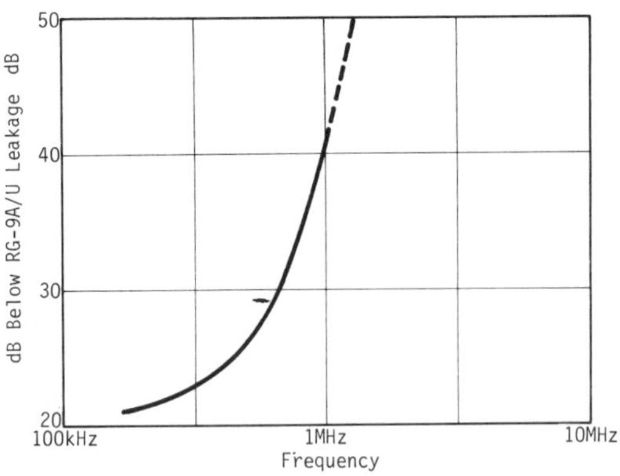

Figure 6.4a - Relative Shielding Effectiveness of Solid Copper Sheath Coaxial Cable. Outer diameter of Inner Conductor = 0.1 cm. Outer Diameter of Outer Conductor = 0.36 cm, and Outer Conductor Wall Thickness = 0.088 cm.

Whenever rigid conduits can be used as inlets of an enclosure, the ordinary commercial galvanized-iron type is found to have the best cost-effectiveness from the viewpoint of shielding. Such pipes and associated fittings are usually heavy enough to give attenuation on the order of 100 dB throughout the frequency range of concern for EMP. In addition, the pipes can be welded as required to maintain continuous shielding, and the conductors can be pulled through after welding so that there is no damage from heat. When used outside, the conduit can be buried, and can have the intimate contact with the earth necessary to dissipate induced surface currents.

Although an ideal shielding enclosure should be continuous throughout and should have no holes, such a situation is rarely encountered in practice. For example, in addition to the inlets and outlets necessary for the flow of signals to and from the enclosure and the power leads, it is often necessary to provide holes in an enclosure for ventilation or moisture drainage. To

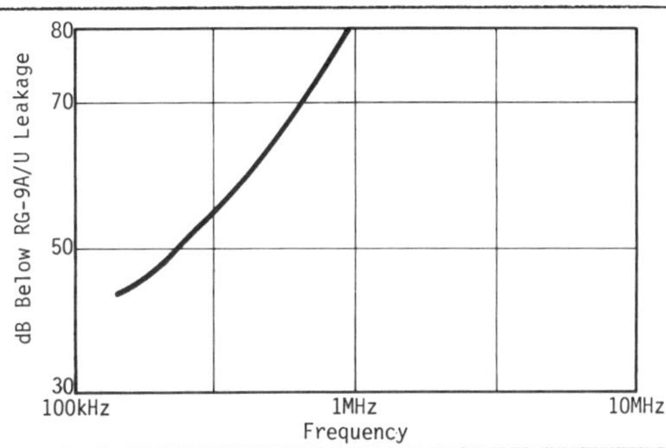

Figure 6.4b - Relative Shielding Effectiveness of Solid Aluminum Sheath Coaxial Cable. Outer Diameter of Inner Conductor = 0.55 cm, outer Diameter Of Outer Conductor = 1.58 cm, outer Conductor Wall Thickness = 0.11 cm.

maintain effective shielding, such holes have to be small in dimension compared with the wavelength corresponding to EMP frequencies of concern. If the holes or apertures on the enclosure wall have to be larger for operational reasons, they have to be covered with a fine mesh of copper screen. When the mesh is easily removable, it should be attached with enough screws or bolts to maintain continuous electrical contact completely around the edge.

The attenuation obtained when an aperture is covered with a screen depends on the size of the screen relative to the cabinet or enclosure, the mesh or perforation dimensions relative to that of the enclosure, and the impedance and distribution of exciting fields. An analysis of a single-hole leakage indicates that for a given exciting field, the leakage intensity at a given distance from the hole is proportional to the cube of the hole dimensions. Typical shielding effectiveness of various meshes and hexcell honeycomb-type screening materials is shown in Figure 6.5. It may be noted that hexcell electroformed meshes have excellent shielding effectiveness, whereas commercial screening materials and perforated sheets appear to be inferior shielding materials.

The steps necessary to make an enclosure shielding-effective include the provision of

1. welded seam construction,
2. serrated spring contact fingers for lids or access doors,
3. waveguide attenuators for control shafts,
4. filtered power leads,
5. screens under dial masks, etc.

Figure 6.6 shows, by way of summary, these necessary features for a shielded equipment case.

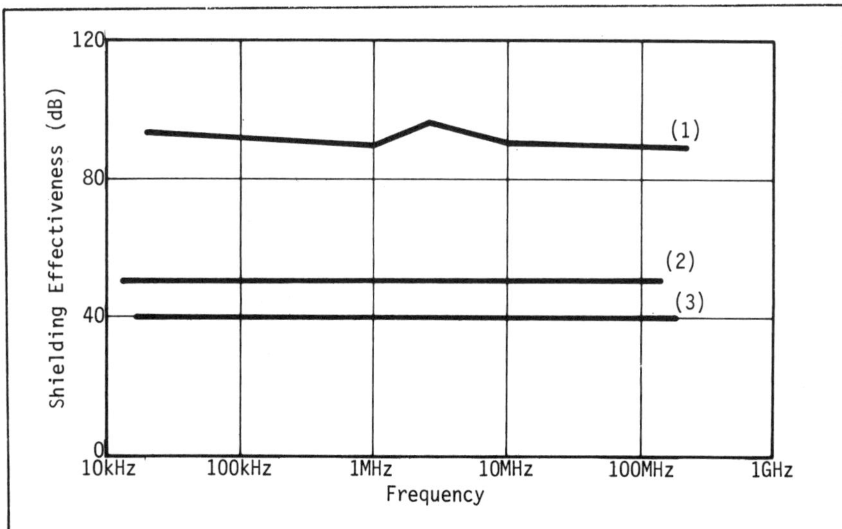

Figure 6.5 - Shielding Effectiveness Of Various Meshes and Aluminum Hexcell Honeycomb Type Screening Materials. (1) 0.64cm Aluminum Hexcell Honeycomb, (2) #16 Aluminum Mesh, 36% Open, 0.02 Wire, (3) #10 MONEL Mesh, 0.018 Wire.

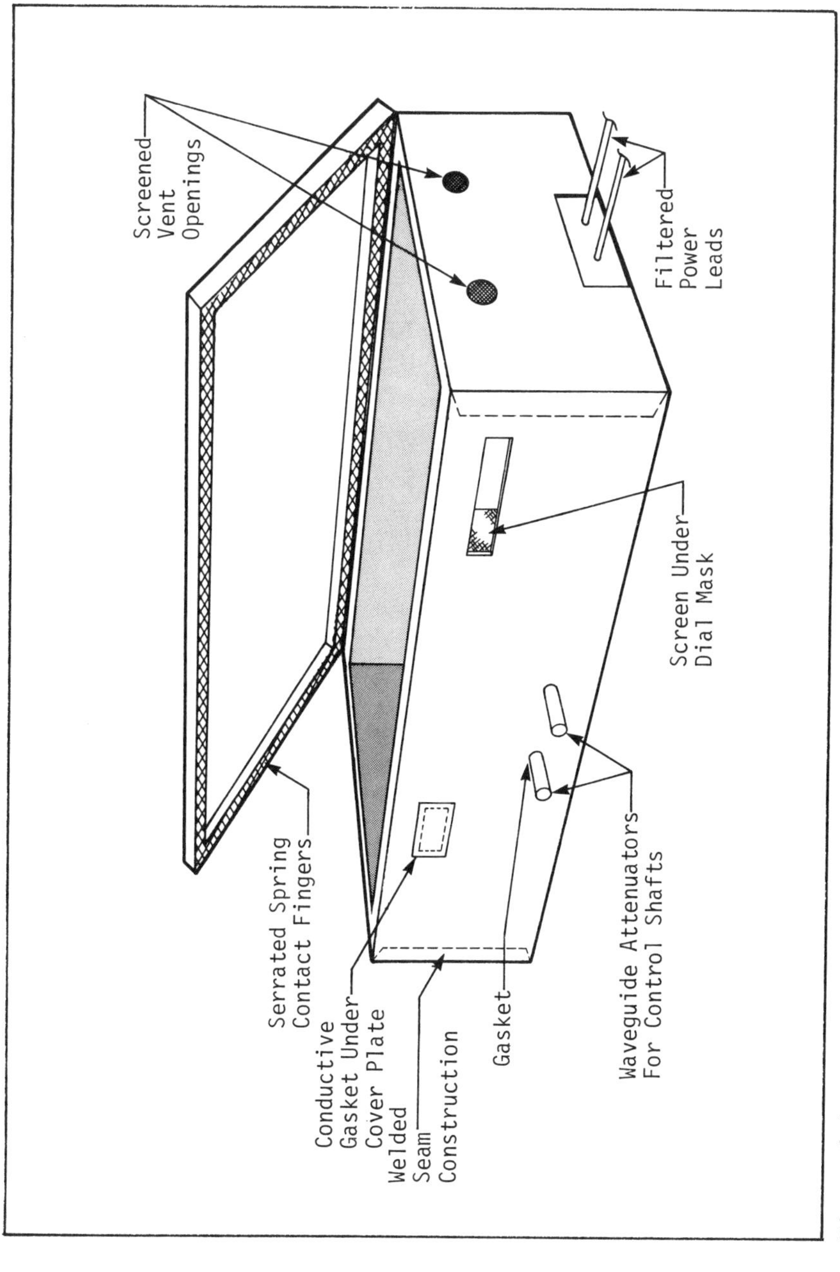

Figure 6.6 - Necessary Features For An Effective Shielded Equipment Case.

6.2 Filtering

EMP coupling into an electronic or weapon system can be reduced by an effective filtering, particularly when the desired signal does not occupy the same frequency band as the EMP. A filter accomplishes this reduction in coupling by discriminating against the frequencies of the undesired signal, which is EMP in this case. The filter, of course, is of little use when EMP energy content in the pass-band of the system is much larger than that of the desired signal, since EMP coupling reduction by means of frequency discrimination is not possible for such a case.

The frequencies in EMP depend on its waveform. An expected spectral distribution of the E-field from a high-altitude nuclear explosion was considered in Chapter 2 (Figure 2.13). Although the spectral distribution is different for the ground burst, it is seen that almost all EMPs are characterized by very high frequency components arising from the initial rise-time of EMP. Such frequencies could be hazardous for most electronic systems, and since the desired signals for the system usually contains frequency components in the lower frequency band, low-pass filters will be more suited for EMP-coupling reduction than high or band-pass filters. Band-pass filters, however, will be desirable in narrow-band communication or radar systems, where the energy content of most of the EMP spectrum outside the pass-band can be substantially attenuated by the filter.

Lumped-element filters for EMP could be T or Π type, as shown in Figures 6.7a and b. Low-pass Π type filters are not desirable for EMP because high voltages can develop across the input capacitor, resulting in degradation or failure under conditions where the capacitor resonates at the source frequency. Hence, T-section filters are usually recommended. Sometimes such filters become ineffective, particularly when the generator and load capacitances resonate with filter inductances L_1 and L_2, thereby providing little attenuation, if any, of the undesired signal. This problem may be avoided by employing ferrite-loaded chokes in the series arm of the low-pass

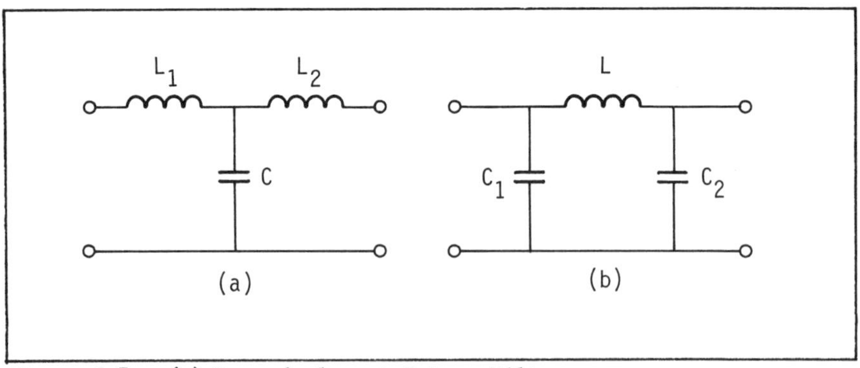

Figure 6.7 - (a) Lumped-element T-type Filter.
(b) Lumped-element π-type Filter.

filters. Such chokes could be designed so that there is essentially no resistance or reactance in the pass-band, but a considerable reactance in the reject band. Ferrite chokes are non-linear and their inductances usually decrease as current increases. This feature makes it difficult to resonate the input or output circuits of the filter because the inductance changes when the current builds up, preventing a high-current resonant condition.

In designing an EMP filter, one cannot ordinarily use classical equations involving interactive and image impedances because the source and load impedances are often poorly defined in many systems, and because they vary widely for the EMP frequency band of concern. In an existing system which requires EMP hardening, measurements of source and load impedances can be made for the frequency band of concern, and appropriate values of such impedances may be chosen for the design of the filter. The filter effectiveness can then be evaluated when the source and load impedances differ at certain frequencies from the chosen values for the filter design. For a new system, the source and load impedances can, perhaps, be properly designed to enhance the effectiveness of the filter.

6.2.1 Typical Filter Responses Due to EMP

Assuming that the source and load impedances of a circuit to be protected against an EMP-induced voltage or current are sufficiently defined so that one can design a low-pass or a high-pass or a band-pass filter, one can evaluate the filtered output and the insertion loss due to the filter by the conventional analysis. Thus, for example, if $V_o(t)$ and $V_i(t)$ denote respectively the time-domain expressions for the input and output voltages of the filter:

$$V_o(s) = T_v(s)\, V_i(s) \tag{6.2}$$

where $V_o(s)$ and $V_i(s)$ are respectively the Laplace-transforms of $V_o(t)$ and $V_i(t)$,defined as:

$$V_o(s) = \int_{-\infty}^{\infty} V_o(t) e^{-st} dt, \text{ etc.} \tag{6.3}$$

and $T_v(s)$ is the voltage transfer function of the filter.

Similarly, when $I_i(t)$ and $I_o(t)$ denote respectively the time-domain expressions for the input and output currents of the filter, one may write:

$$I_o(s) = T_I(s) I_i(s)$$

where $I_o(s)$ and $I_i(s)$ are the Laplace transforms of $I_o(t)$ and $I_i(t)$ respectively and $T_I(s)$ is the current transfer function of the filters.

For a Butterworth type low-pass and high-pass filter,

$$T_V(s) = \tfrac{1}{2}\,\frac{\omega_c^3}{(s+\omega_c)(s^2+\omega_c s+\omega_c^2)} \quad \text{for a low-pass filter}$$

$$= \tfrac{1}{2}\,\frac{s^3}{(s+\omega_c)(s^2+\omega_c s+\omega_c^2)} \quad \text{for a high-pass filter} \qquad (6.4)$$

where, $\omega_c = 2\pi$ (cutoff frequency)

The corresponding expressions for the current transfer functions are:

$$T_I(s) = \tfrac{1}{2}\,\frac{\omega_c^2}{(s+\frac{\omega_c^2}{2}) + (\frac{\omega_c}{2})^2} \quad \text{for a low-pass filter}$$

$$= \tfrac{1}{2}\,\frac{s^2}{(s+\omega_c)^2 + \omega_c^2} \quad \text{for a high-pass filter} \qquad (6.5)$$

For the band-pass Butterworth filters, the expressions for the transfer functions are:

$$T_V(s) = \tfrac{1}{2}\,\frac{1}{(x+1)(x^2+x+1)}$$

and,

$$T_I(s) = \tfrac{1}{2}\,\frac{1}{(x+\frac{1}{2})^2 + \frac{1}{4}} \qquad (6.6)$$

where, $x = \frac{\omega_o}{B}\left(\frac{s}{\omega_o} + \frac{\omega_o}{s}\right)$

$$\omega_o^2 = \omega_{c1} \times \omega_{c2}$$

$$B = \omega_{c2} - \omega_{c1}$$

ω_{c1} = 2π x (lower cutoff frequency)

ω_{c2} = 2π x (higher cutoff frequency)

Let it be assumed that the electric field due to EMP impinging on a dipole antenna of effective height h_e can be expressed as:

$$E(t) = A\,e^{-\alpha t} + B\,e^{-\beta t} \text{ volts/meter, for } t \geq 0$$
$$= 0 \text{ for } t < 0 \qquad (6.7)$$

where A, B, α and β are arbitrary constants* and the E-field is parallel to the dipole. The voltage impressed at the output of the antenna, then, becomes:

$$V(t) = h_e \left[A\,e^{-\alpha t} + B\,e^{-\beta t} \right]$$
$$V(s) = h_e \left[\frac{A}{\alpha + s} + \frac{B}{\beta + s} \right] \qquad (6.8)$$

If a low-pass filter is used at the antenna line to reduce EMP coupling to the system, from Equations 6.5 and 6.9 one obtains:

$$V_o(s) = \frac{h_e}{2} \frac{\omega_c^3}{(s + \omega_c)(s^2 + \omega_c s + \omega_c^2)} \left[\frac{A}{s + \alpha} + \frac{B}{s + \beta} \right] \qquad (6.9)$$

The time-domain expression $V_o(t)$ for the filtered output may be obtained from the inverse Laplace transform of Equation 6.1. If the antenna is a small loop of radius r with an inductance L, and if the impinging EMP magnetic field is:

$$H(t) = Ce^{-\gamma t} + De^{-\delta t}, \qquad (6.10)$$

the current induced in the loop due to EMP may be approximated as:

$$i(t) = \frac{H}{L} \pi r^2$$

---

* This waveform corresponds to EMP from high altitude burst as shown in Figure 2.12, where B = -A.

and,

$$I(s) = \frac{\pi}{L} r^2 \left[\frac{C}{\gamma + s} + \frac{D}{\delta + s}\right] \quad (6.11)$$

From Equations 6.6 and 6.12, the low-pass filter output becomes:

$$I_o(s) = \frac{\pi}{2L} r^2 \frac{\omega_c^2}{(s + \frac{\omega_c}{2})^2 + (\frac{\omega_c}{2})^2} \left[\frac{C}{\gamma + s} + \frac{D}{\delta + s}\right] \quad (6.12)$$

Again, the time-domain response, $i_o(t)$ for the filtered output may be obtained from the inverse Laplace transform of Equation 6.13.

It is evident from Equations 6.10 and 6.13 that the transient response of the filter and the insertion loss, implying an absorption or reflection of EMP energy by the filter, will depend on the values of ω_c and α and β or γ and δ. Also, since α, β, γ and δ cannot be controlled for the filter design, the only parameter available to reduce EMP coupling by a low-pass filter is the angular cutoff frequency ω_c.

It is possible in many cases to design low-pass, high-pass or band-pass filters to provide an insertion loss approaching 100 percent for that portion of the frequency band over which EMP energy is distributed. However, even for an insertion loss above 95 percent, a substantial amount of energy, on the order of tens of joules or higher, may reach the protected circuit, whereas the typical damage level may be on the order of tenths of a joule. Moreover, in cases where a filter prevents damaging amounts of energy from reaching the protected circuit, this energy must eventually be absorbed either by the filter itself or by resistive elements preceding the filter. Therefore, a filter which effectively protects a load may itself fail or cause some other portion of the circuit to suffer damage. Thus, the primary consideration for the EMP filter design must be to provide means to absorb or dissipate the EMP energy from which the load or a circuit has to be protected. Lossy inductances, or the use of powdered iron as a core material for the inductances, may be considered as filter elements for this purpose.

Since EMP coupling into a system by power lines usually constitutes a vulnerable area, considerations to prevent such couplings by appropriate filters shall be in order. Most ac power lines do not provide adequate protection against EMP transients. Although ac power filters are normally expected to exceed 100 dB attenuation from 14 KHz to 1 GHz in a 50 ohm system, many installations have voltage gains at various frequencies in the 1-20 KHz region. EMP filters for ac power lines should employ inductive inputs and capacitive outputs. The inductor assures the firing potential for the preceding arrestor and limits the current through the filter capacitor. The input inductor should be so wound, and the start and finish loads should be so separated, that the inductor could withstand a high voltage breakdown test. The capacitors in the filter should be of a self-healing type.

Dc line filters should have similar attenuation characteristics and should be preceded by spark- or gas-discharge arrestors. Dc lines may also be protected by transient bucking transformers. This scheme is especially effective in the protection of dc-to-dc converters from burnout due to collector-emitter voltage resulting from line transients.

Often dc circuits employ semi-conductor regulators or other protective circuits which are susceptible to line transients. Similarly, the presence of polarized capacitors in dc lines could have severe consequences when any reverse transients occur due to EMP. This type of susceptibility suggests the use of special power interrupting or quenching circuits.

6.3 Voltage Limiters

Another means to reduce EMP coupling into electronic or weapon systems is to employ voltage limiters at strategic points. Such a limiting is done by Zener diodes, nonlinear resistive devices, arc breakdown protectors, etc.

Zener diodes can be used to protect circuits against voltage levels significantly higher than signal voltages. They have the property of passing electric current easily in one direction at a low voltage, but inhibiting the current flow in the other direction until Zener voltage is exceeded. This characteristic makes the Zener diode useful as a voltage limiter since for the desired signal voltage, less than the critical voltage, the Zener diode acts as an open circuit, whereas the same diode acts as a short circuit for voltages above the critical voltage. From the viewpoint of protection agains EMP, typical applications of Zener diodes include power-supply transient suppression and protection of circuits which may be sensitive to voltages above a critical level. Zener diodes may also be series-connected to achieve a higher breakdown voltage, and they may be used back-to-back to protect electronic circuits subject to both positive and negative overvoltages.

The impedance characteristic of various Zener diodes with different Zener voltages is shown in Figure 6.8. Controlled avalanche Zeners have extremely high-power handling capability for short surges. Typical surge capacity of small Zeners is shown in Figure 6.9. Some Zener diodes have essentially instant response. It is this characteristic which is very useful to prevent large transients due to EMP from propagating to sensitive electronic circuits. However, the fast switching characteristics of Zeners may cause undesirable transients unless provision is made for dissipation or absorption of such transients.

Voltage-limiting to reduce EMP coupling can also be accomplished by a variable resistance (usually shunted to ground), the value of which depends on the voltage or current. Such characteristics are exhibited by Varistor or Thyrite type materials. The variable resistance is useful as a protection device, because an increase of voltage and current causes the resistance-to-ground at the circuit input to decrease. The approximate relation for the voltage and current in such cases can be expressed as:

$$I = KV^N \text{ Amps} \tag{6.13}$$

where I is the current, V is the voltage across the resistor, and K is a constant depending on the resistor material, and is typically on the order of 10^{-10}. The value of N also depends on the resistivity of the material, a typical value of N being about 5 for a resistivity ρ = 10,000 ohm-cm at 28 volts. When N = 5, for example, a doubling of the voltage V increases the current by a factor of 32. In addition to offering a lower effective resistance with the increasing voltage, Varistors resist damages due to EMP energy by withstanding a large amount of energy, on the order of 2 to 3 thousand joules per cubic inch volume of the resistor material.

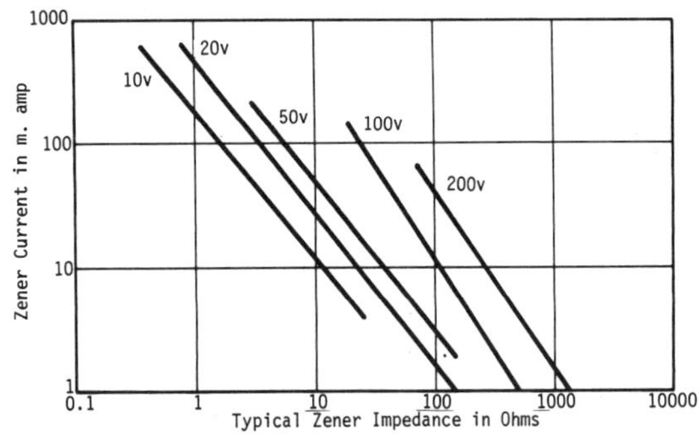

Figure 6.8 - Typical Zener Impedance As A Function of Zener Current Of A 5-Watt Zener.

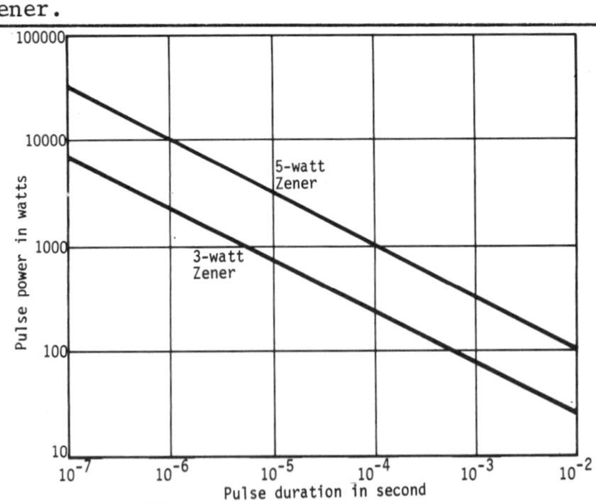

Figure 6.9 - Typical Surge Capacity Of Small Zeners. Square Pulse Power vs. Non-repetitive Pulse Duration.

The material most commonly used for Varistors, Thyrite, etc., is silicon carbide. Silicon carbides exhibit a symmetrical nonlinear resistance with an exponential voltage-current relation like, or similar to, that shown in Equation 6.14. Silicon carbides are also electrically stable and can respond to fast microsecond impulses the same way as they respond to ac or dc signals. A typical mode of application of a nonlinear resistance for the over-voltage protection purpose is to insert a tightly fitted silicon carbide annular ring or disc between the inner and outer conductor of the coaxial cable which could couple EMP energy into an electronic or weapon system.

General Electric Company* has introduced a metal-oxide type Varistor which obeys the nonlinear Equation 6.14, has a small size, nanosecond switching speed, and is capable of handling hundreds of amperes of surge currents. The construction and characteristics of such a Varistor are shown in Figure 6.10a.

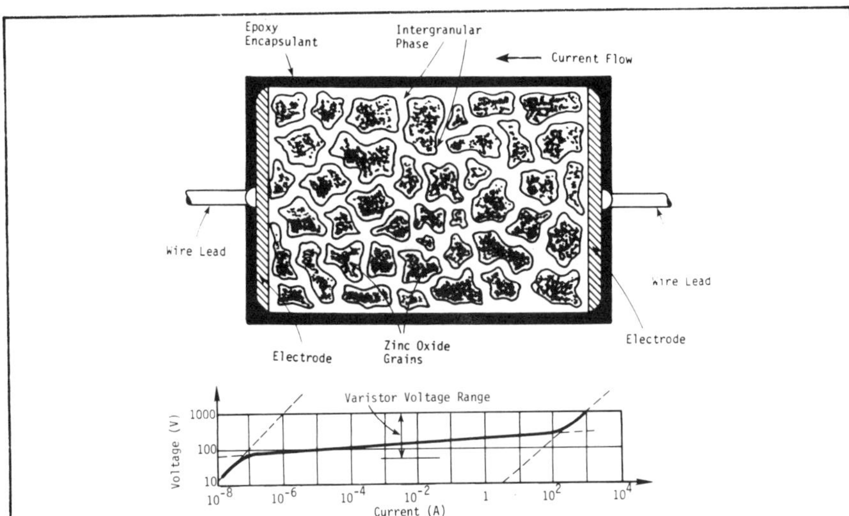

Figure 6.10a - Example of MOV as an EMP Protective Device.

Over-voltage limiting to prevent excessive EMP energy coupling into a system can also be achieved by a gas-breakdown protector. Typically such protectors consist of a pressurized gas in a chamber formed by an air-core coaxial line and two dielectric plugs in the form of annular rings placed at the two ends of the chamber as shown in Figure 6.10b. The kind of breakdown voltage V that could be achieved between concentric cylinders in air is estimated as:

$$V_b \simeq 31pa_1 \ \ell n \frac{a_2}{a_1} \left(1 + \frac{0.308}{\sqrt{pR_1}}\right) \qquad (6.14)$$

* J. D. Harnden, Jr., and F. D. Martzloff, "Metal-oxide Varistor: A New Way to Suppress Transients", Electronics, October 9, 1972.

where p is the air pressure (atmosphere), a_1 and a_2 are respectively the inner and outer conductor radii in cm and V_b is the static breakdown voltage in KV. However, if the inner and outer conductors are supported or sealed by the dielectric plugs as referred to above, the breakdown will generally initiate with a dielectric-stimulated arc which usually forms at a potential lower than the value corresponding to air breakdown. The voltage-limiting device with dielectric stimulated arc breakdown responds to fast transients corresponding to EMPs with a very short rise-time. Thus, in some cases, it may be desirable to eliminate the gas chamber and allow the breakdown to take place across the dielectric in open air.

The breakdown voltage for this type of protection device can be reduced by using neon or helium or neon-argon mixtures instead of air and by using gases at a lower pressure. Since gas leaks can seriously degrade the performance, care must be taken to design gas-breakdown type protective devices accordingly.

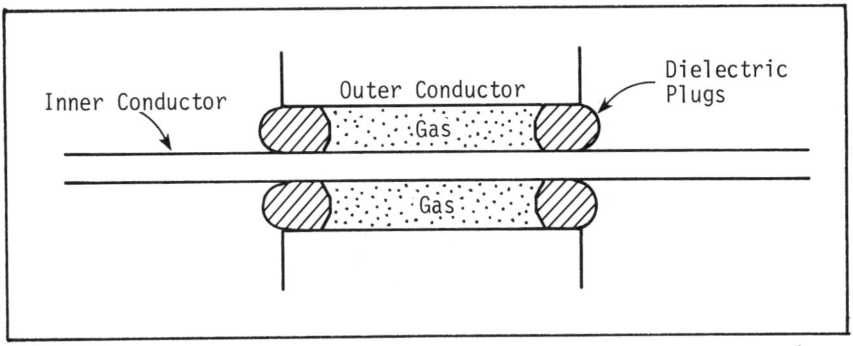

Figure 6.10b - Pressurized Gas Chamber Around An Air-core Coaxial Line As A voltage-limiting Device.

Because of the switching action of the gas discharge device, the energy dissipated within the device is minimal. The balance of the energy contained in a transient is dissipated in other resistive components of the circuit including circuit wiring, as well as resistive components of inductors and capacitors.

With Zener diodes, Thyrites, and Varistors, however, the energy of the transient must be dissipated within the bulk materials of the surge arrestor. The electrical energy is converted into heat, and the heating effect is normally destructive to the operation of the device. Table 6.1 shows a comparison of some characteristic features of Zeners, Varistors and gas-discharge devices. Table 6.2 shows the operating voltage ranges and response time for some commonly used voltage limiting devices.

Another silicon device, Transzorb*, introduced by General Semiconductor Industries, may be used for EMP transient suppression. It contains a large area pn junction having integral heat-sinks, and is capable of handling short-duration high-power pulses (1,500 watts for 1 millisecond, 100,000 watts for 100 nanoseconds).

Any voltage limitor, as noted above, seldom constitutes a complete solution by itself insofar as the protection of circuits or subsystems against EMP is concerned, because each has some limitation in speed of response and range of surge voltage. Thus, EMP protection devices are usually hybrid, making use of a combination of different types of protection concepts.

TABLE 6.1

Comparison of Voltage Protection Devices

Device	Capacitance Farad	Peak Current Capability (Amps)	Pulse Energy Dissipation (Jules)	Temperature Range Capability (Degree C)
Zeners	10^{-8} - 10^{-9}	10 - 10^{2}	10^{-2} - 10	-40 to 85
Varistors	10^{-9} - 10^{-10}	10^{2} - 10^{3}	10 - 10^{2}	-40 to 85
Uni-Imp**	10^{-11} - 10^{-12}	10^{3} - 10^{4}	10^{2} - 10^{3}	-55 to 125

* O. M. Clark, "EMP Transient Suppression Using the Transzorb," General Semiconductor Industries, Inc., Appl. Note No. 1012, Tempe, Arizona, 1979.

** Uni-Imp devices are gas discharge devices introduced by Signalite/Division, General Instrument Corp.

TABLE 6-2

Typical Response Times of Protection Devices

Device	Operating Voltage Range (Typical)	Response Time (Typical)
Zener	5V to 20KV (dc)	10^{-4} to 10^{-7} Second
Gas Breakdown	100V to 1000 KV (dc)	10^{-1} to 10^{-7} Second

6.4 Network Hardening

Protection against EMP can be effected to a limited extent by appropriate circuit or network designs. In such cases, a logical design ensures protection against arbitrary transients through coding or timing discrimination or both. For example, a sequential system can be designed so that a particular sequence of pulses according to a code, instead of a single pulse or a stray transient, is necessary for a logical state change. EMP, which is like a stray single pulse, cannot, therefore, effect a state change and hence introduce error and a consequent system failure. In some cases, a two-input flip-flop circuit, in which a pulse at one input causes the flip-flop to set to "one", while a pulse at the other input causes the flip-flop to reset to "zero" (simultaneous pulses at both inputs being necessary to cause the flip-flop to trigger or effect a state change), can be used to provide protection against EMP.

The protection against EMP can also be effected by time discrimination in network response. Thus, if a network is designed so that the desired signal must be present for a particular period of time which is longer than that likely to be due to EMP, one can discriminate against EMP transient and thus avoid any possible error due to EMP. In some situations, discriminations can also be provided by parity techniques where a network response, by design, will require a signal to appear on two differently coded lines. Unlike the desired signal, an arbitrary transient such as the one due to EMP, is unlikely to satisfy the parity requirement and hence introduce an error.

6.5 Device Hardening

In examining a system to ascertain its EMP vulnerability for a given threat, it is sometimes found that the hardening of a particular device or component of the system may significantly improve overall system hardness. Often a cable or cable connectors or antenna input circuit or a shield or a grounding system can be redesigned, replaced or modified relatively easily to make an electronic system invulnerable to a specified EMP threat. Sometimes, a device hardening may involve selecting high voltage levels for transient immunity, screening potentially weak metallization and bonds, voltage or current limiting across critical components, redundant construction, etc. Although device-hardening is often straightforward, care must be taken during this process so that other components are not unduly weakened or overburdened.

6.6 Circumvention

Circumvention involves sensing an EMP field that could upset or damage some sensitive components or the system, and discontinuing signal processing until the danger of transient malfunction disappears. Circumvention is feasible since the damaging EMP transients are often of short duration from the viewpoint of normal operation of the system. If EMP and desired signals, whose levels are high enough to cause logic changes, are detected, a circumvention may be effected by blocking all inputs and recycling the sequence of operations to the last safe reference point. After a designated time-delay corresponding to EMP waveform, operation resumes by resetting where needed and updating to account for the elapsed time. A second type of circumvention uses parallel signal processing by an extremely hard but slow digital signal processor which does not respond to fast EMP transients and a soft, fast processor. Ordinarily, the system is under control of the fast processor. If and when EMP is detected, the operation is temporarily turned over to the hard processor, which subsequently resets the fast processor, enabling normal signal processing when EMP transients are gone. The third type of circumvention involves normal operation which turns a warning light when EMP is detected, so that an operator can reset the system when the hazardous EMP transients no longer exist.

A circumvention can also be effected by providing two parallel paths in the input circuit, one for the desired signal and the other operative only when the signal level exceeds a predesignated value, which is higher than that of the desired signal, but much lower than the EMP-induced transient level. As illustrated in Figure 6.11, the difference of the signals in two paths constitutes the new input to the system. In the absence of EMP, this difference signal is the same as the desired signal since no signal flows through the second path. As EMP strikes, the transients flow through two equal paths cancelling each other when the subtraction of signals is made.

An effective circumvention can also be obtained in some cases by using a ferrite-core transformer. Since a saturation of the core takes place when the time integral of input voltage ($\int Vdt$) exceeds a certain value, the core can be designed so that there is no saturation for the desired signal level. When the large EMP-induced transients at the transformer primary exceed the limiting value of $\int Vdt$, the core-saturation will limit the level of the secondary voltage, thus providing a protection against EMP as long as the transients last. A normal operation is restored when EMP transients disappear.

Sometimes it may be necessary to combine a number of protective and circumventive devices or techniques to obtain the desired degree of protection against EMP. An example of such a use is illustrated in Figure 6.12.

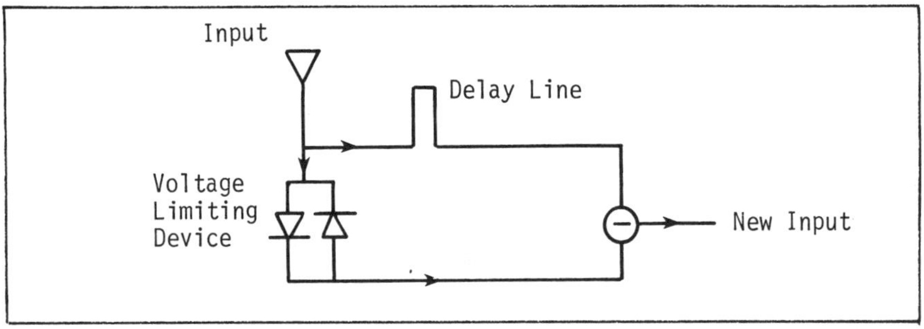

Figure 6.11 - A Circumventing Technique Where EMP Transients Are Blanked Out From The System Input.

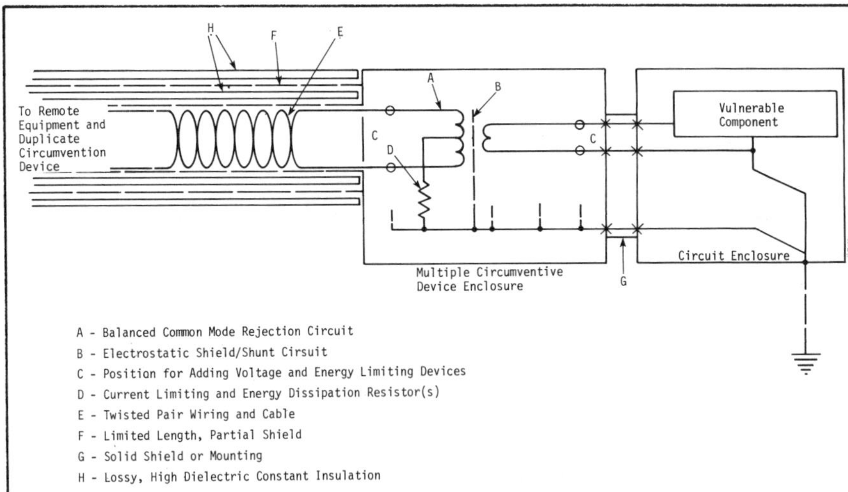

Figure 6.12 - Use of Multiple Circumvention Devices And Techniques.

6.7 Good Design Practices as Hardness Tools

One of the most powerful means of protection against EMP is to comply with good design practices for the system. Generally, an engineering design effort deals initially with the satisfaction of the necessary capabilities and functions of a system, and then with the imposition of various requirements for durability, invulnerability, maintenance, etc. The two aspects are not symmetric, in that such trade-offs as may arise, will generally favor the "offensive" or mission-accomplishing features over the "defensive" or survival features. It is possible, however, to formulate and comply with a set of design rules which strike a balance between the offensive and defensive purposes.

The impact of design rules are not exactly the same for the initial design and retrofit measures. For providing EMP hardness in existing systems, design rules may suggest fixes after violations of such rules are identified. One needs to formulate a general set of rules which will minimize vulnerability for existing systems and will be a guide to detect violations for existing systems so that fixes can be introduced. The same set of rules will be a guide for the design of new systems.

It is to be recognized at the outset that no design rules can reduce EMP threat or change functional characteristics of a system. Also, engineering judgments have to be excercised to apply the rules in any system.

From the viewpoint of electronic or weapon system design, the design rules should be applicable for the following areas:

a. Circuit Layout
b. Shielding
c. Cabling
d. Grounding.

6.7.1 Design Rules for Circuit Layout

For circuit layout, design rules based on "protective orientation" and "voluntary criteria" could be as follows:

1. Establish a common, centrally-located "Reference Node Structure" for each cluster. This is the "low side" collection of circuit connections and lines, including "chassis ground", "circuit ground", "neutrals", "shields", "racks", etc. Reference Node may be a large "grounding" strip or a metal lug panel.

2. Group a single circuit in a single cluster or zone. Avoid circuits having sections in different nodal areas.

3. Establish a complex of cluster node relative to the reference node - this is the collection of "signal hot" connections between identifiably individual elements or zones of a system. System or cluster node may be plug array or a terminal strip mounted on the reference node.

4. Connect all elements to one another through the system node complex. There should be only one such path between any two components.

5. For complex systems, more than one node structure may be essential. In such cases, connect nodes by means of single trunk cables arranged in a "tree" or "radial" configuration. All circuit connections must pass through that trunk, including "grounds" or reference node inter-ties.

6. For some critical features, make system interconnections only at a "node" array located as close as possible to the reference node. There should be no point-to-point wiring and no "grounds" at far end of reference node wiring.

7. Connect a (cluster) conductor to a cluster or reference node only once.

8. Group all conductors as twisted cables, each cable containing one, and only one, complete circuit.

9. Provide low impedance common reference paths independently for circuits, and for major shield components, isolated from exterior "ground" everywhere except at intentional tie points.

10. All system and intrasystem wiring should conform to a "tree" (one main trunk and branches) or radial (spokes) layout scheme. This includes all conduits, cables, power, signal, communication and ground routes, as well as non-electrical conductors such as fuel lines, water pipes, mechanical control shafts, cables, etc. It should also apply to all subsystem wiring of low-impedance or high-sensitivity circuits, within cabinets, chassis, or racks.

11. Where possible, lay out the system elements or circuit components along the cable trunk in order of "noise" level and component susceptibility.

12. Avoid loop layouts, particularly for low-impedance circuits.

13. Avoid circuit routes and paths which couple to other circuits.

14. Install cables and wires in electrically continuous conduits or "cope trays".

15. Provide independent, insulated return (low-side) conductors on all signal and power cables. This is especially important in circumstances requiring "common mode" suppression. Do not use wire and cable shields as low-impedance paths or current conductors for functional circuits between system or circuit components. Provide separate low-impedance return conductors. Isolate "common" side from shield "ground."

16. Use shielded wire for high-impedance circuits. Do not use shield or circuit "low-side" return. Provide independent conductors within the shields.

17. Locate high-energy or high-sensitivity circuits (elements and wiring) in central areas away from shield corners or major joints. Avoid running wires and cables through system extremities.

18. In examining a system, identify all possible conductors and conducting paths. Check for "weird or hidden ones" such as:

    Fuel Lines
    Fire Control
    Swear Lines
    Railroad Tracks
    Elevator and Hoist
    Cables, tubes, rails, etc.
    Traffic Control
    TV Monitors (closed circuit)
    Photography
    Covert Microphone Systems
    Special Circuits, e.g. documentation

19. Exceptions to reference node system:

    a. Power supply circuits remotely located. But use twisted and shielded "non-loop", internally complete cable system.

    b. High-speed, low-capacitance circuits. Use low-impedance input-output design with low-impedance terminated cables, separately shielded and entirely "self-contained" and isolated circuits.

20. Use gated or strobing circuits in digital systems (sensing, control or computing circuits). Choose gate widths which are intermediate between highest and lowest circuit and system response frequencies. Design the operating circuits specific to the gate pulse structure and not to other possible inputs.

21. Avoid "compton diode" geometrics in exposed elements. Design these for minimum or self-cancelling (balanced) induced compton currents.

6.7.2 Design Rules for Shielding

For shielding, design rules, when applicable, should be:

1. Provide complete independent conductive enclosures (including "reentrant circuits" around each zone or bounded region.

2. Shields can be used both to isolate an offending source, as well as to protect sensitive elements. Therefore, shield all critical areas.

3. Use several stages of shielding, if necessary, instead of one thick one.

4. Zone interfaces should be separated wherever possible. Avoid common wall shielding between system sections.

5. Keep circuit clusters compact and well within shields. Avoid corner regions.

6. Provide separation between circuit groups to minimize intercoupling of inductive fields.

7. Provide distributed energy dumps between and within shield enclosures such as coupled resistors.

8. Multiple contacts between system conductors should be periodically spaced with random separation. Avoid regularly spaced contacts.

9. Shielding joints should be continuous, low-resistance contacts: continuous weld, braze, sweating, swaging. Next best: aperiodic contacts - many rivets, bolts-star washers, spot welds, continuous solder, conductive gaskets, spring leaf strip, etc.

10. For necessary pressure contacts, surfaces are to be: non-corrosive, clean, bare metal. Avoid dissimilar galvanic contacts. Provide for easy inspection and replacements.

11. Shielding enclosures should not be penetrated by major conducting structural members or service ducts, e.g., steel building beams, or metallic sewer lines. Such elements should be restricted to separate columnar and layered regions.

12. On a small scale, incompletely protected penetrations can include: metal holes, control elements-switches, control shafts, etc.

    Frequently the metal part (switch, handle, shaft) is completely floating and also does not completely "plug" its aperture. Hence, it can act as a electrostatic coupler, a current conductor and an electromagnetic field leak.

13. Avoid shield apertures, keep necessary apertures as small as possible. Cover with screening, electrically bonded around hole.

14. Watch especially "inadvertent" penetrations such as screening solid duct-work, air conditioning outlets, non-metallic screws, etc.

15. Provide automatically closing doors for shielded rooms. These should be recessed, complete, pressure-contact electromagnetic seals and multiple flexible bonds.

16. Specify solid welds at all rebar contacts or crossings.

6.7.3 Design Rules for Cables

Design rules on cabling involve the treatment of intercluster connections. Some useful rules in this regard are as follows:

1. Shield critical cables separately (source power, sensitive circuits).

2. Bury (intersystem) cables as deep as economically feasible (more than a meter). Most earth media with a conductivity of 10^{-2} to 10^{-3} mho/m provide a good skin current damping.

3. Check continuity of cable shield layers. Also, check inter-shield resistance. Splicers often do not connect shielding correctly. Shield layers should not be connected to each other, nor to "ground" except for cases discussed in rules on grounding.

4. System shields should be connected to corresponding cable shields, in proper order, without intercrossings between environmental zones.

5. Avoid ordinary braid shielding. Use solid metal coax, or specially woven, multi-layer braiding.

6. Use double or triple coax, if needed.

7. Install cables in unbroken conduits.

8. Do not interrupt outer cable shielding at tap-in junctions.

9. Provide terminations for each electric path in multi-layer shielded cables (as well as for inner conductors). Terminate each propagation path or duct resistively in its characteristic impedance.

10. Provide reflective interfaces such as large metal sheets perpendicular to cables and lossy (ferromagnetic) couplers around or "in" cables to dampen unwanted propagation modes, such as skin currents.

11. Ensure good contact bonds between conduits (or outer cable shields) and circuit enclosures or intermediate boxes. Avoid use of non-conducting lubricants when preparing and assembling conduits.

12. If cable path loops are unavoidable, provide shield-damping elements for removing circulating currents.

13. Preferential low-resistance paths can be provided by means of close-coupled "bus" wires laid alongside cables.

6.7.4 Grounding Rules

Another area, where the compliance with design rules can significantly minimize EMP coupling and susceptibility problems in a complex system, is grounding. Generally, it deals with the "low-potential side" of circuits. Some design rules useful for grounding are as follows:

1. Provide a single-point "ground" for a circuit cluster, usually at the biggest piece of the system electrically immersed in the earth, such as the water supply system.

2. The "outer" reference node or zone should be extensively "earth" grounded. This may be achieved by intimate burial, or by frequent periodic grounding rods.

3. All "outer" reference elements should be thoroughly and frequently bonded to one another. This includes outer enclosures, cases and shields, conduits, metallic services (water, fuel, sewage, etc.).

4. If "outer" reference contains intrinsic loops, such as in a rectangular building framework, these may be treated by well-bonded "closures", i.e., metal skins or screens connected all the way around.

5. Include all non-essential conductors and equipment such as water pipes, elevators cables, etc., in the outer reference system. These should not penetrate through the more sensitive shielded inner zone.

6. Conduits, cable shields and other metallic penetrations should be "well-bonded" at enclosure entries and exists.

7. Don't rely on "pigtails" for bonding. Bond straps should be wide, short and thin.

8. Intermediate reference shields, cable layers, and circuit busses should usually be single-point "grounded" at one location in a given system cluster. However, ground bond for one reference level should not penetrate through the next.

9. Do not carry circuit grounds through common paths with other circuit grounds or with shield reference grounds. Provide separate low-impedance bonds from one level to the next.

10. Avoid circuit ground paths through multiple separated bonds (or ground loops).

11. Avoid parallel grounding conductors which follow different paths to the same points, thus closing a loop.

12. Metal enclosures containing high-energy sources or very sensitive circuits should be insulated from the "earth" ground. Generally, such enclosures should be surrounded by the outer reference shield which will be intimately "earthed".

13. Eliminate all "ground" loops in circuits, particularly phantom loops through "balanced" systems such as ac neutrals, Y-Y systems, etc.

14. Individual circuit chassis and racks should be insulated from "outer" reference node, i.e., structural framework. Bring back ground through separate low-resistance wire paths in cable conduit. Internal ground wires should be insulated.

15. Do not carry circuit "grounds" through the same electrical path as chassis/shield ground wires and bonds. Also, provide independent circuit return leads.

16. "Ground" the common circuit reference at one end only, usually at the reference node point.

17. Neutral and center points should be grounded at one end only. Avoid long ground leads.

18. Extended systems usually require separate "grounding" at each circuit cluster, especially when such clusters are far apart.

19. Long connecting cables and conduits should be treated by damping elements and by frequent outer-layer grounding.

20. For long cables or shields, "ground" to earth or to major structural reference at approximately every 1/6 of a wavelength at the principal frequency, but locate such points with a random spacing.

The primary objective of the precautions on grounding, as outlined above through the design rules, is to avoid unnecessary radiation and excitation of large currents in ground loops when EMP impinges on such loops. Another objective is to avoid unnecessary intercoupling between different zones, circuit clusters, etc.

6.7.5 Design Rules for Protective Steps

In addition, a set of design rules may be formulated to provide protective steps to overcome design shortcomings in some cases, particularly from the viewpoint of EMP vulnerability. They are:

1. Where complete shielding is not possible, pass intrasystem circuits (cables) through a large high-impedance reflection plane. This sheet or box should be thoroughly bonded to the outer cable shield or conduit.

2. Use bifilar chokes and unity transformers to interrupt unavoidable cable loops, common mode paths, phantom circuits, etc.

3. Use shielded transformer coupling to avoid long common leads or loops.

4. All non-signal penetrations such as power lines, control lines, TP wiring, etc., should be filtered at the point of entry, preferably using feed-through type units.

5. RFI line filters should be shielded independently and individually. Do not place the entire burden on a single line filter.

6. Provide a single-phase surge protection on each phase of a multi-plan power system. Use fast-acting circuit breakers rather than fuses.

7. Apply "lightning protection" on all above surface lines.

8. Provide surge protection on the emergency power system, high-voltage systems and oil-filled transformer. Use dry air-cooled transformers wherever possible since oil-filled transformers can explode and spread burning oil.

6.8 References

1. *Electromagnetic Pulse Handbook for Missiles and Aircraft in Flight EMP Interaction 1-1*, Air Force Weapons Laboratory, September 1972.

2. Morrison, R., *Grounding and Shielding Techniques in Instrumentation*, Wiley and Sons, 1967.

3. Ferber, R. R., and Young, F. J., *Enhancement of EMP Shielding by Ferromagnetic Saturation*, IEEE Trans. on Nuclear Science, Vol. NS-17, No. 6, December 1970.

4. Hays, J. B. and Bodle, D. W., *Electrical Protection of Tactical Communications Systems, Final Report*, Bell Telephone Labs, Inc., August 1968.

5. Uhlig, E. R., OCE NEMP Program, *Development of Criteria for Protection of NIKE-X Power Plant and Facilities Electrical Systems Against Nuclear Electromagnetic Pulse Effects: Protective Measures*, General Electric Company, December 1967.

6. Brown, G. L., Ghose, R. N., Green, J. X., and Wouters, L. F., *Electromagnetic Pulse Energy Coupling Studies*, ANC 30R-11, Vol. III, American Nucleonics Corporation, Glendale, 1967.

7. Wik, M., Kapp, W. H., Eggendorfer, A., Johl, W., Buchmann, W., *Measurement and Application of Secondary Surge Arresters for the Purpose of NEMP Protection*, Fourth Symp. and Tech. Exhibition on EMC, Zurich, March 1981.

8. *EMP Protective Systems*, Defense Civil Preparedness Agency, TR-61-B, July 1976.

9. *EMP Protection for AM Radio Broadcast Stations (Glossary of Design Practices)*, Defense Civil Preparedness Agency, TR-61-C, July 1976.

10. Vance, E. F., *EMP Hardness of Systems*, Fourth Symp. and Tech. Exhibition on Electromagnetic Compatibility, Zurich, March 1981.

11. Clark, O. M., *EMP Transient Suppression Using the Transzorb*, General Semiconductor Industries, Inc., App. Note 1012, Tempe, Arizona, 1979.

12. White, D. R. J., *A Handbook on Electromagnetic Shielding Materials and Performance*, Don White Consultants, Inc., 1975.

CHAPTER 7

EMP Hardness Assurance Program

To avoid loss of function or capacity of military systems due to EMP, a hardness assurance program should be carried out for every system which must function during or after a nuclear strike. For a new system, this program should begin with the system acquisition phase and extend to its entire life cycle, accommodating any and all engineering changes in the system and its retrofits. To be effective, the hardness assurance program should not only consist of the application of a body of technology but also a management which will ensure system hardness in a planned procedural manner, by qualified personnel, encompassing design, development, protection and operation of the system. Further discussions on the purpose, scope, plan and execution of the hardness assurance program suitable for electronic or weapon systems and a methodology for the formulation of standards and specifications to be used for system acquisition are outlined in this chapter.

7.1 Hardness Assurance Program

EMP hardness assurance program is a set of activities which should be added to various existing functions in the life cycle of a system to define and control its EMP hardness. The purpose of conducting a hardness assurance program is to provide and maintain, in a cost-effective manner, a level of hardness or invulnerability for the system which is commensurate with its survivability requirements in an EMP environment. For cost effectiveness, the hardness assurance program should be initiated upon the decision to acquire a new system, and should remain in effect until the requirement for survivability is withdrawn. In the case of a system already in the definition, development or production phase, the hardness assurance program should be started upon the issuance of the EMP survivability requirement.

Principal activities of the hardness assurance program should include:

- Determination of the hardness level requirement and assessment of the need for a hardness assurance program for a particular system.
- Assignment of priorities, responsibilities, schedules and criteria for the creation and maintenance of hardness.
- Collection, collation and generation of susceptibility data related to sensitive components, subassemblies, or the system.
- Assurance that all system aspects which may affect hardness are systematically evaluated and controlled.
- Consideration and resolution of problems involving hardness maintenance due to expected and probable changes throughout the life cycle of the hardened equipment.
- Formulation of test objectives and concepts for hardness assurance.
- Formulation of standards, specifications, technical requirements and procedures.
- Development of a plan for educating appropriate personnel in hardness assurance disciplines.

To determine the hardness level requirement, systems may be classified on the basis of the region where they are located (such as ship, ground base, aircraft, missiles, etc.), on the basis of their functions (such as communications, radar, computer, weapon systems, etc.), and on the basis of operating frequencies, bandwidth, etc., since EMP impact on the system depends on them. Then, from the threat scenario, a document is developed, describing the possible and probable EMP environments in terms of field strength, waveform, angle-of-arrival, polarization, etc. Also, at this time, an EMP simulation facility, which may be used to test the system's survivability potential, may be selected.

Concurrently, during the preparation of this data, the cognizant agency or personnel of the user should prepare an EMP impact document describing in some detail the predicted (electromagnetic) EM effects, and outlining the minimum protective designs required to control these effects. These design requirements are to be added to the end item description of the system. From the operational requirements, EMP environment, and EMP impact document, one may decide the type of hardness assurance program, if any, which will be appropriate for the system of concern.

7.1.1 Formulation of Hardness Assurance Program

Once the decision to employ a hardness assurance program is made, one of the principal tasks to be undertaken in formulating the program will be to define and delineate authorities and responsibilities of personnel or organizations to enforce the program. For an existing system, this may involve either creating a new organization specifically for the purpose or appointing an administrative hardness engineer with appropriate authorities and responsibilities to enforce EMP hardness doctrines. Also, since EMP hardness is not the only requirement in a system, priorities have to be arranged to ensure that other system requirements are not compromised or unduly undermined while providing EMP hardness. For a new system, the appropriation of responsibilities and authorities should be effected before the system is acquired. If the new system is to be designed or built by a contractor, for example, the user may have to assume the responsibility for preparing the operational doctrines and functional requirements for the system, EMP impact document and appropriate specifications, and for selecting or generating applicable standards. The contractor will, in this case, be responsible to design, fabricate and test in accordance with the specification, and to provide documentation showing various EMP hardness measures undertaken for the system, including EMP invulnerability verification test results, when and as appropriate.

A hardened system, in general, requires hardened components and subsystems. A certain volume of data is presently in existence concerning the susceptibility of electronic equipments and parts due to EMP-induced voltages and currents. The collection of existing data will reduce the magnitude of the susceptibility test program. For an existing system whose hardness is of concern, the susceptibility data and tolerance thresholds for various semiconducting devices and other sensitive circuit components should be collected or generated, if they are not available. Similar information should also be collected or generated for a new system by one responsible for the design of the system. For the hardness assurance, the sensitive components and subassemblies have to be protected from EMP influence, either by reducing EMP coupling into such components or subassemblies by shielding, filtering, etc., or by replacing them with more hardened components or subsystems. In any event, a baseline susceptibility data for all components and subassemblies of the system of concern should exist before any remedial measure may be taken for hardening.

One of the key elements in hardness assurance of a system, whether existing or yet to be built, is a systematic evaluation of all aspects of the system. Although as a practical matter, actual hardening may be required for a few components or subassemblies, in general, the selection of components or subsystems which need attention can only be found after all aspects of the system have been evaluated. For this purpose, one may have to perform coupling analysis, determine circuit performance and enclosure shielding, and evaluate protective devices, cables, connectors, etc. For existing systems, it may be necessary to control some aspects of the system operation, if permissible, to avoid the problem, instead of evaluating the outcome without such control. Similar controls can also be effected in a new system at the design stage.

Finally, it should be recognized that the hardness assurance is not a one-time task which is required only when a system is designed or when EMP hardness requirement is first imposed on the system. Instead, it must extend to the entire life cycle of the system. Many systems, for example, undergo changes at various stages of their life cycles because of obsolesence of some components or subsystems or because of technological advancements or changes in operational needs, all of which require engineering changes or retrofits. Means must be provided to assure hardness when such changes take place in the system. Since such means may not likely be available in any standard or specification used during the acquisition phase, an organization must exist to control or implement these operations. Standards and specifications themselves may have to be modified to accommodate the changes, or other plans and procedures must be formulated for this purpose. In any event, a separate hardness management organization with appropriate responsibility must exist to administer hardness assurance functions during both the acquisition and the system life cycle.

7.1.2 Hardness Assurance Tests

An effective EMP hardness assurance program could hardly be formulated without adequate test provisions which verify the degree of expected invulnerability. Since it is not practical to conduct any tests in an actual nuclear-induced environment, EMP invulnerability verification tests are carried out in simulation facilities. The development of test objectives, selection of the required simulation facility, formulation of specific test or tests, and the data analysis and interpretation of test data, therefore, constitute key tasks in a hardness assurance program. Available facilities which can simulate various ranges of EMP threats, or those which could be designed for specific test purposes, are further discussed in the following section. The selection of a simulation facility should be made in the system definition phase for a new system. For an existing system, such a selection should be made when the threat and the required degree of mission-oriented invulnerability are defined.

For hardness assurance purposes, tests are often conducted for diagnostic and remedial reasons, instead of hardness certification on a "go-no-go" basis, although such a certification may sometimes be required. Thus, initial planning and execution of simulation tests are carried out to discover unacceptable deviations from the predicted system response when subjected to a specified EMP threat. Such tests are followed by processing and analyzing simulation data. This may involve extrapolation of system response for threat level when the simulation is effected in a low-level facility. Fourier analysis, operation of computer circuit models, comparisons of threat with susceptibility, etc., are the usual means available for this purpose.

Following this data processing, a "first" vulnerability status report is prepared which includes a statement of vulnerability status of the system as designed, together with the data, analysis, and extrapolations from which the vulnerability statement is derived. A discussion of confidence level of the test and statement is also to be included in this report.

This report usually suggests design modification and fixes which will upgrade system invulnerability to the required levels. Hardware modification based on the report is then incorporated into the system, often on a "breadboard" basis at first. The hardware need only be electromagnetically equivalent for this purpose. If the modification is acceptable (both necessary and sufficient), then prototypes are prepared and tested. A "second" report follows this test, containing the evidence that the system has been hardened in the most cost-effective manner. Confidence level factors should also be included in the report. Finally, the engineering design test is carried out to prove that the hardening modification has not affected the overall system performance. EMP hardness certification, if required, is issued following such a proof.

7.1.3 EMP Protection Provision in System Specifications

Since systems and equipments are designed, tested and maintained in accordance with designated specifications and standards, it is imperative from the viewpoint of EMP hardness that such specifications and standards must contain appropriate EMP provisions. Specifications particularly addressing EMP hardness problems should contain:

1. List of specific EMP design elements
2. List of EMP protective devices
3. Test criteria
4. Suggested test methods
5. Performance criteria before and after EMP strike

6. Susceptible elements and circuits
7. A section containing threat data.

While formulating or modifying specifications and standards for EMP hardness, one should note the similarities between the EMP and other electromagnetic interferences (EMI) which may affect the system performance. For example, electronic and weapon systems, which are not subjected to EMP threat could be designed, nevertheless, in accordance with specified EMI and electromagnetic compatibility (EMC) requirements. Such requirements define electromagnetic threats, specify design guidelines and tests. When EMP hardness requirement is superimposed on those for EMI/EMC, it is often realized that the design criteria for one set of specified threats are common to those for the other set of threats. Similarly, tests required to be conducted to prove system immunity against one set of threats could also satisfy the test requirements for other threats. Good design practices followed for reducing EMP vulnerability of a system often remove unwanted EMI impact on the system, and tests conducted to verify satisfactory compliance of EMI requirements may sometimes be adequate for EMP hardness assurance. Specifications and standards, therefore, could and should be formulated to consolidate the same or similar electromagnetic interference protective features of the system and to avoid unnecessary tests for the invulnerability verification. Thus, although the formulation of appropriate specifications, standards, and technical requirements are essential elements of EMP hardness assurance program, it should not be treated as an isolated problem. The concept of development of standards and specifications which not only address the EMI/EMC requirements, but also EMP requirements, are further discussed in Section 7.3.

7.1.4 Maintenance of EMP Hardness Assurance

The maintenance of EMP hardness assurance program for a system will require that strict disciplines be followed. Because of the unusual character of such disciplines, it will be necessary to develop an educational or training program for responsible personnel who will maintain the system hardness during its entire life cycle. Trained operational personnel, who should be certified accordingly, must be proficient with protective devices, RFI connectors, shielded enclosures, cable systems and components whose susceptibility may be enhanced if improperly used. In general, providing training and general maintenance support will be the responsibility of the system user. Trained personnel will also be needed by the user for acceptance tests (both performance and documentation). In addition, a field training program will be required in some cases to ensure that the system integrity is not affected from the viewpoint of EMP hardness during routine and unusual field maintenance.

Typical trainee classifications for EMP hardness program may include:

1. Management officials who will determine operational doctrine, deployment or logistic criteria, do war gaming, etc.

2. Scientists, who will work in a largely theoretical area and will be concerned with prediction, explanations of observations, design of non-conventional tests, and development of new concepts.

3. System and subsystem design and development engineers and technicians, test engineers and technical support personnel.

4. Manufacturing engineers, foremen, quality assurance engineers, value engineers, reliability engineers and production test engineers.

To maintain the integrity of the hardness assurance program, qualified personnel should be certified, and only certified personnel should be entrusted with hardness assurance responsibilities.

Many of the elements of hardness assurance program as discussed above evolve from good engineering practices and a system management sense. These elements are by no means all-inclusive, and often some of them will have to be deleted as they are inappropriate. Similarly, new elements in the hardness assurance program may have to be added to suit specific systems or specific circumstances.

7.2 Methodology for EM Standards and Specification Formulation

For the assurance of hardness of a system which may be subjected to EMP threat, appropriate standards and specifications must exist so that equipments acquired and maintained in accordance with such standards and specifications will withstand specified threats, without unduly degrading the system performance. Although the hardness assurance problems are not confined to the acquisition phase of the equipments, and usually extend to the entire life cycle of the equipments, one is concerned with standards and specifications mostly when equipments are acquired for the first time. Also, standards and specifications must address all the system problems rather than those involving only EMP threats and consequential remedies.

When a system is required to withstand a specified EMP threat, at least a part of the standards or specifications used for system acquisition must be devoted to EMP threats and system requirements to overcome such threats. Many systems which are to be protected against EMP, however, are also required to be protected against other electromagnetic (EM) interferences such as those originating from power lines, other systems, electronic counter-measure signals, etc. Often the same systems are also required to be compatible with other systems, such that electromagnetic leakages from the systems in the form of radiation, conduction, etc., are kept below specified levels. Standards and specifications to be used for a system, therefore, must include provisions necessary to protect it against all specified EM threats, including EMP, and to maintain appropriate compatibility requirements with other systems.

As stated earlier, there could be a set of common design criteria for many EM threats. Also, the same set of tests could be used to verify system immunity against a number of, if not all, EM threats. From the viewpoint of cost effectiveness, such commonalities should be utilized while formulating standards and specifications to be used for the system acquisition phase. Thus, one may hope to formulate EM standards and specifications as a whole, instead of addressing separately EMP, ECM and other EM compatibility problems. The formulation of such EM standards and specifications could be best achieved by a planned methodology. This methodology will thus be a "road map", starting with the objective of electromagnetic hardness, and leading to the details of specifications and standards with plans and procedures to achieve and maintain hardness. Steps involved in this methodology may include:

1. Objectives
2. Scope
3. Background and existing data
4. Assessment of electromagnetic (EM) threats
5. Assessment of systems
6. Requirement analysis
7. Applicable standards and specifications for the acquisition phase
8. Assessment of existing standards and specifications and their deficiencies
9. Formulation of required standards and specifications
10. Formulation of rationale for exception to standards and specifications
11. Formulation and maintenance of various data bases
12. Test provisions.

7.2.1 Objectives

For the formulation of standards and specifications, which will be the essential tools to assure system invulnerability, the principal objective may be stated as follows:

Assurance of system invulnerability from all expected environmental EM effects so as to maintain an operational and readiness capability of the electronic or weapon system of concern. Such environmental EM effects will include the presence and consequence of all probable, natural, friendly and hostile electromagnetic threats and their future growths.

7.2.2 Scope

The scope of the standards and specifications will be all electronic or weapon systems which may be subjected to EM threats and which may be affected as a consequence.

7.2.3 Background and Existing Data

There are numerous EM standards and specifications available now for various aspects of EM threats, such as EM compatibility, EM vulnerability, EM nuclear pulse, EM safety, etc. While some deficiencies exist in many of the available standards and specifications, and while some threats such as electronic countermeasures (ECM) and EMP are not adequately addressed in any such standards and specifications, one could make use of this background as the starting point for the formulation of an all-encompassing or separate but complementary EM standards and specifications. Such a use will avoid unnecessary duplication of efforts.

7.2.4 Assessment of EM Threats

EM threats of concern may be classified as those causing EM incompatibility, vulnerability, and safety hazards, and EMP, ECM and power-line interferences. For the purpose of the formulation of any EM standard or specification, these threats must be viewed from the characteristics of their sources, frequency band, spectrum, intensity and their synergistic effects, if any. The organization to acquire such characteristics of EM threat data is regarded as an essential step for any requirement analysis for the formulation of EM standards and specifications. Also, for this purpose a data base needs to be established providing appropriate threat parameters. Means must also be provided to update this data base and to accommodate system changes and threat growths.

7.2.5 Assessment of Systems

Specifications and standards required to assure system invulnerability against specified EM threats must evolve from the interaction of the threat and the system of concern, and the resultant system susceptibility. Consequently, an assessment and classification of systems must precede the formulation of such standards and specifications. For this purpose, systems may be classified on the basis of:

1. Regions where they are located or installed
2. Functions
3. Operating characteristics such as frequencies, instantaneous bandwidths, etc.

Examples of classifications of systems based on regions are:

- Ship
- Shore or ground-base
- Aircraft
- Submarine
- Satellite

Such a classification is important since threat and performance requirements of the same equipment located in different regions are, in general, different.

Examples of the classification of systems based on their functional characteristics are:

- Communication, command and control network
- Radar
- Navigation and navigational aids
- Missile guidance and targeting
- Computers

Classification of systems on the basis of functional characteristics is important since system-upset or vulnerability depends to a large extent on its functional requirements.

To assess the interaction of threats and systems, one needs to classify the systems or equipments also on the basis of their operating frequency, bandwidth, etc. Examples of such classifications are:

- EHF
- UHF and LF
- VHF and UHF
- Microwaves ranging from C and D bands to J and K bands

Since various threats have distinct spectral characteristics, and an interaction of the system and threat occurs, in general, only when the threat frequency spectrum coincides with the frequency spectrum of the receiver of concern, a classification on the basis of frequency and bandwidth becomes necessary to formulate the requirements for any standards and specifications.

7.2.6 Requirement Analysis

Requirements for the formulation of standards and specifications, which will be needed to assure system hardness against EM threats, follow directly from the system characteristics and threats to which the system is subjected. Figure 7.1 shows a schematic diagram of a mathematical model which can be used for the requirement analysis. In this figure, one set of inputs consists of the characteristics of the systems. In the model, these characteristics are allowed to be operated by the threat parameters. There could be several threats, such as EMI, EMP, ECM, etc.

Each threat needs to be characterized on the basis of its sources, frequency bands, spectrum, etc., as discussed earlier. The requirement criteria for the hardness assurance follow directly as an output of the model. Since, in general, no parameters other than those of the system and threat characteristics are involved to determine system susceptibility or vulnerability, the simple model, shown in Figure 7.1, may constitute an adequate basis for the formulation of requirements for EM standards and specifications. It is recognized that although the model is conceptually simple, the operation of the model to derive standard or specification requirements as output must be supported by adequate data base or bases, which will characterize the systems and the threats.

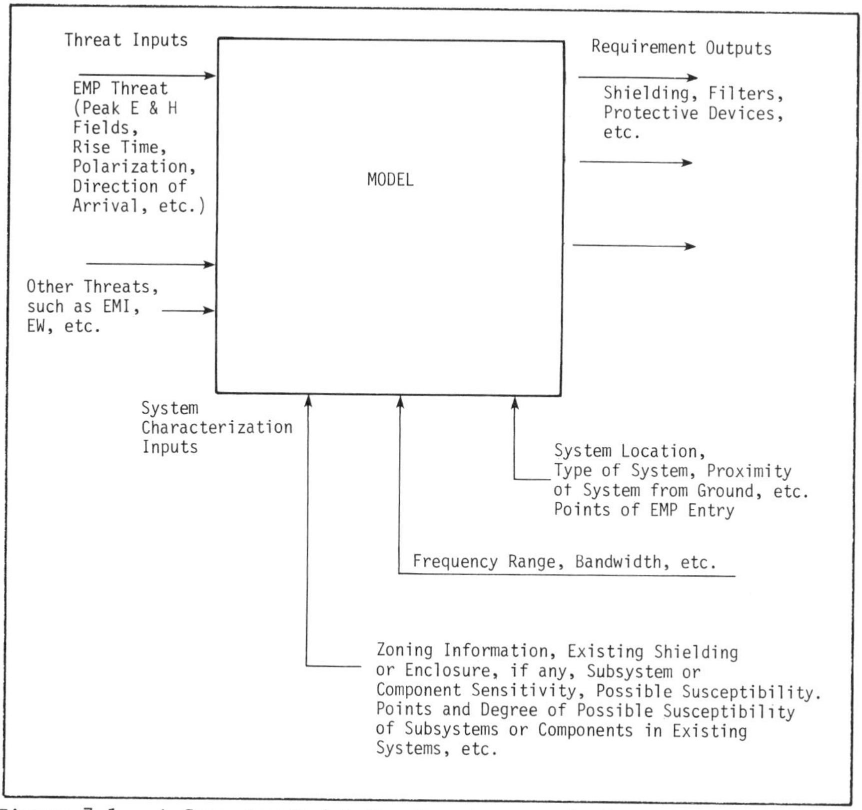

Figure 7.1 - A Computer Model For Requirement Analysis.

The direct output of the model under consideration will provide functional requirement data for the formulation of standards and specifications. A typical example of a functional model that may be used for the formulation of EMP specification is shown in Figure 7.2. Details of the EMP specification follow directly from the functional requirement tree as shown in the figure. Similar requirement tables can be formulated for other threats such as ECM, etc.

Functional requirements shown in Figure 7.2 are qualitative in form. Since no formal specifications can be formulated with qualitative information alone, the purpose of exercising the model, as discussed earlier, will be to generate corresponding quantitative parameters.

7.2.7 Applicable Standards and Specifications for Acquisition

The output of the model, however, should not be restricted only to the generation of requirement tables. More specifically, one needs to know exceptions to specifications and standards which must be made during the acquisition of a particular system, so that unnecessary cost burden is not imposed. Any such exception, however, has to be examined carefully to ensure that true requirements to achieve and maintain system invulnerability are not compromised willfully or unintentionally through exceptions. Following such an examination, one may be able to determine the necessary elements which will be applicable to standards and specifications to be used during the acquisition phase.

7.2.8 Assessment of Existing Standards and Specifications

Before formulating new EM standards and specifications for acquisition, it is desirable to assess existing standards and specifications prepared for similar systems or equipments. Many provisions in existing standards have resulted from experience gained over a long period on good design practices, on cost-effective means to circumvent EM interference effects in systems, and on appropriate tests to verify system objectives. A comparison of the elements of applicable standards and specifications derived from the requirement analysis, as discussed earlier, with the provisions of existing standards or specifications for similar equipment or systems, will be helpful toward drafting the new standard and specifications. The text and language of common elements, for example, could be lifted directly or with minor modifications from existing standards or specifications and set in the new ones. Similarly, an examination of the differences will assure avoidance of unnecessary provisions in new standards or specifications, or the introduction of provisions which should be included in new standards and specifications but omitted at the output of the requirement analysis as discussed earlier.

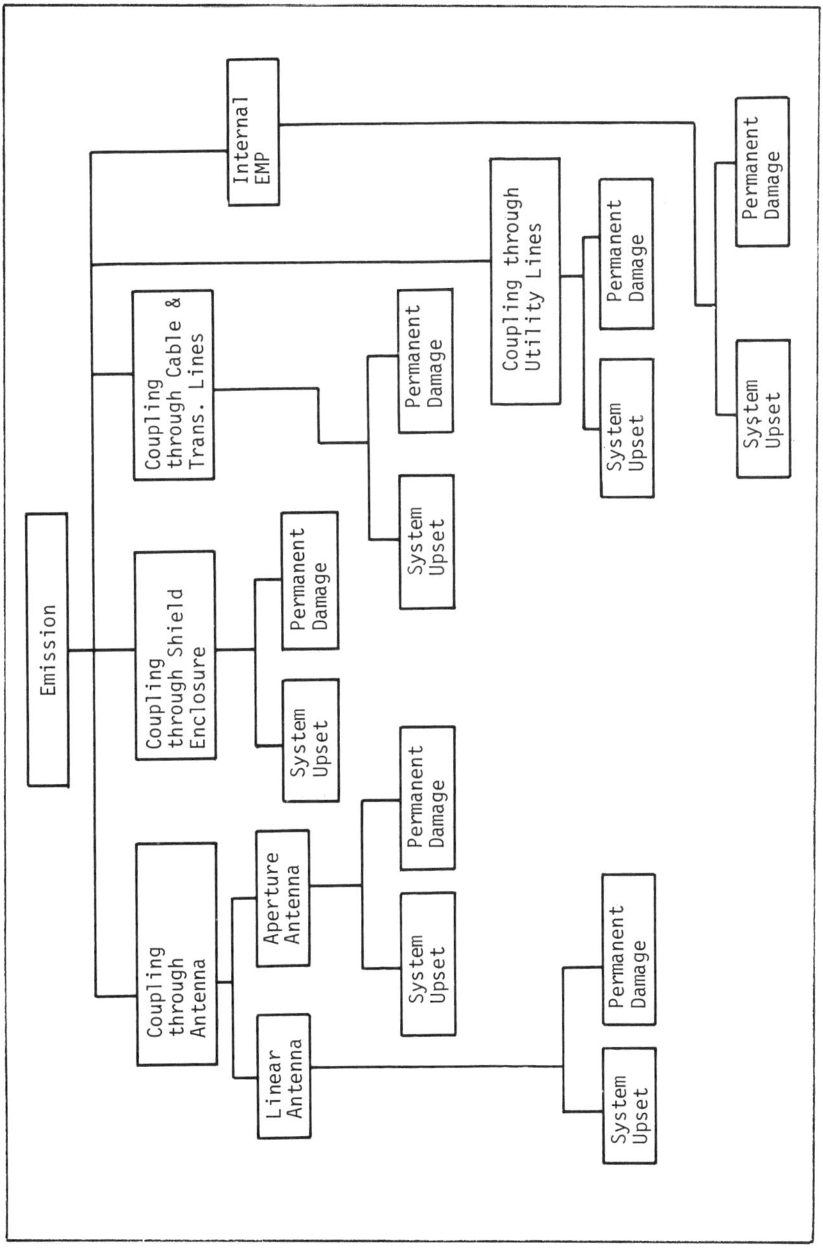

Figure 7.2 - An Example Of A Functional Model For Formulation Of EMP Specifications.

7.2.9 Formulation of Required Standards and Specifications

The requirement table, as noted in this section, could be so arranged as to provide input to the specific language to be used for the draft of standards and specifications. The task of the formulation of a standard or specification to be used for a particular equipment or system during its acquisition, then, becomes that of the formulation of appropriate technical and administrative directives which address each of the blocks shown in the requirement table. Since the requirement table should provide both qualitative and quantitative aspects of technical and administrative directives, a complete standard specification could be obtained directly from such a table.

To ensure the EM hardness of the system, various hardness-assurance tests will have to be performed on the system, such that the compliance of the desired test results will be indicative of the suitability of the system for acquisition. Various tests required to meet with the hardness assurance standards, however, have to be specifically designed so that test results are directly related to the hardness objectives in a cost-effective sense. It is likely that the design of such tests may not follow exclusively from the requirement table. Additional engineering input will, therefore, be required at this phase to complete the standard or specification.

7.2.10 Formulation of Rationale for Exception to Standards and Specifications

Although a standard or specification calling for anything less than the required degree of hardness of the system of concern is unacceptable, it is recognized that any excess in a standard or specification incorporated mostly for the purpose of standardization should be equally unacceptable in most cases because of the unnecessary cost implication. Since some standards are necessary in specifications for the case of management of hardness assurance for multitude of systems, means must be provided to effect exception to any provision of the specification when needed. As mentioned earlier, enforcing exceptions to any specification is a sensitive issue, since a misuse of such a flexibility, intentionally or unintentionally, will negate the basic hardness assurance objective. To ensure a balance, plans and procedures to effect exceptions must be set as an integral part of the hardness assurance program, and they should be regarded as just as important as the standards and specifications for which they are intended.

7.2.11 Formulation and Maintenance of Various Data Bases

The data pertaining to the system, including component sensitivity and susceptibility, and EM threats to which the system is likely to be subjected, constitutes a key element for the successful implementation of a hardness assurance program. Similar data is also essential for the formulation of standards and specifications to be used during system acquisition. The data on tolerance, thresholds and burn-out voltages and currents for various semiconducting devices, which are frequently used in electronic or weapon systems, for example, should constitute a data base. Similarly, characteristics of various shielding enclosures, EMP filters, and protective devices should also be a part of the data base pertaining to systems. This data base should be updated when new developments or an obsolescence suggests such updating. Various types of threats as applicable to radars, communications and computers located in aircraft, ships, etc., should also be included in the data base. For the convenience of management of various data, one may maintain a number of different data bases, instead of an all-encompassing one. Such bases may, for example, be on:

a. Systems based on their regions of installation, function, operating frequency, bandwidths, etc.

b. Component sensitivity susceptibility, etc.

c. Threats

d. Existing standards and specifications and their effectiveness

The primary objective of these data bases will be not only to constitute starting points for the formulation of standards and specifications, but also to provide guidelines for the system design, accommodation of engineering changes in the system, etc.

7.2.12 Test Provisions

To ensure system hardness during its acquisiion, various hardness assuring tests will have to be performed on the system, such that a verification of the desired test results will be indicative of the suitability of the system for acquisition. Various tests required to meet with hardness assurance criteria, however, have to be specifically designed, so that test results are directly related to hardness objectives in a cost-effective sense. The design of such tests may not follow exclusively from the requirement table, and additional engineering input will be required at this phase to complete the specification or standard.

For many EM threats, such as EMI, test provisions in existing standards and specifications should be examined to determine whether such provisions could be adapted directly or with modifications for new EM standards and specifications to be used for the acquisition of a system. Since the threat magnitudes are much greater for EMP in comparison with those for other threats, special tests and test facilities are required. Test objectives and simulation facilities which could be used to assure EMP invulnerability of an equipment or system during its acquisition phase are further discussed in Chapter 8.

7.3 Hardness Assurance Maintenance and Test Set

Perhaps implied in the hardness assurance program discussed in Section 7.1 is a hardness assurance maintenance program, since hardness requirements usually extend to the entire life cycle of a system. Many aspects of such a maintenance program have already been discussed in Section 7.1. Additional elements of the hardness assurance maintenance program should include hardness assurance test set, various manuals and a training program.

Hardness assurance test set may be a portable test set which can be connected to an existing system, or it may be a built-in device, integrated with a system which makes hardness assurance tests routinely or when commanded. The development of a hardness assurance test set begins with a set of diagnostic tests to demonstrate the measurable effects of various likely and possible faults. The data analysis of these tests will determine the best simulation source, its type (cw or pulse), field-strength, rise-time and waveshape, and expected voltage and current levels at selective probe-points. The design concepts can then be decided, and a prototype of a test set may be designed and built. Since a single data base may not lead to a worst-case or optimum design for the test set, a parallel program to devlop a data base on a good statistical basis should be initiated. The hardness assurance test set may have to be made adjustable if the data indicates such a need.

While the test set prototype is being tested, the installation and field maintenance procedures may be prepared. These will eventually become part of the hardness assurance test set manual and the hardness assurance maintenance program manual.

Hardness assurance maintenance program manual is a general document, not associated with any particular electronic or weapon system test set or test procedure. It should be tutorial in nature and should describe cleaning, fitting, replacements, adjustments, and other processes and techniques associated with hardness maintenance. Some of the subject headings of the manual will be grounding, bonding, metal surface treatment, gaskets, finger stock, riveting, welding, soldering, connections, cable assembly, filter mounting, feed-through devices, enclosure door hardware, etc.

The hardness assurance maintenance program should include the preparation of a general test manual which contains detailed instruction on the use of hardness assurance test set, preparation of test items, test procedures, reporting forms and instructions, test schedules, test-crew requirements, test set maintenance, calibration, standards, follow-up, etc.

7.3.1 Training Program

The hardness assurance maintenance program should also include a factory training program designed to train scientists, engineers, and factory personnel, including factory field teams. Other special training programs are required for operators and maintenance technicians.

Subject matters for the courses taught for the factory training program should include, as a minimum, electronics - EMP source, covering nature of the threat, comparison of EMP with other noise sources, EMP energy penetration paths, failure modes and mechanisms, protective devices, test equipment and procedures, repairs and replacements, hardness assurance test sets, trouble shooting, connector maintenance, etc. Similarly, the subject matter for the operator training program should include, as a minimum, grounding precautions, failure indications and their meanings, fundamentals of threats and their effects, use of software, changes in operating procedures, etc.

7.4 Retrofit EMP Hardening of Existing Systems

The hardness assurance program and hardness assurance maintenance program as discussed in previous sections are primarily administrative programs for ensuring and maintaining EMP invulnerability of electronic or weapon systems. Sometimes, it may be required to provide a retrofit hardening of a system before formulating or adapting any hardness assurance program. Steps to be followed for such a retrofit hardening are somewhat similar to those discussed in Chapter 5 under EMP vulnerability analysis methodology, although the emphasis in this case should be to harden the system instead of predicting system vulnerability. The first step for this hardening operation then becomes a preliminary examination which includes:

- Geographical take-offs of electromagnetic features onto transparent sheets so that different portions of the system may be overlaid and studied for interaction

- Redrawing of pertinent portions of a circuit (the grounding scheme, for example)

- Studies of specifications, process instructions and inspection reports to see whether a finished product should be acceptable

- Disassembly or X-rays of hardware

- Listing of semiconducting devices and cross-checking against their susceptibility criteria.

- Study of circuits (worst-case analysis).

- Study of system action.

Implicit in the preliminary examination routine is the idea of zoning which, along with its other functions, identifies all components or subsystems that should not be in strong fields due to EMP. Once the preliminary examination has been made and hardness violations are identified, it is often possible, by simple tests, to determine the effect of such violations. Such tests usually involve subjecting the system or parts of the system to the field from a simulator. The detecting instruments are then connected to monitor the sensitive circuits. Various EMP energy penetrating paths are covered and uncovered, connected or disconnected, one at a time, so that the specific effect of each step may be observed.

7.4.1 Temporary Shields and Hardness Measures for Breadboard

Very often, major susceptibility problems in a system are avoided by appropriate shielding enclosures and cable shieldings. To determine the extent of hardness fixes obtainable by such enclosures and cable shieldings, one may conduct tests with the simulator referred to above, and temporary shields which can be placed at appropriate places. These temporary shields may be regarded as electromagnetic equivalents of breadboards which are frequently used in electrical engineering to try out design concepts and their effectiveness.

Temporary low-cost electrostatic cable shielding may be obtained by using zippered sheath* metal foils or spare braid. Usually the zippered sheath is a foil or braid in a convenient plastic jacket and has a seam that runs axially along the side under the zipper. If flexibility is required, a metal braid can be slipped over a cable, even when there are large connectors at the ends of the cable. The braid, with a diameter which could be stretched over a ratio of 3.5 to 1, can cover the connectors as well as the cable and is terminated at the end of the cable by hose-clamps at the enclosure walls.

The braid or zippered-sheath type construction for cable shielding is also good for lower-level H-field threats, although copper-plated iron wire should be used as the shielding material for such cases. If this is not good enough and some flexibility is still required, then RFI flexible conduit may be temporarily patched-in so that the conduit is not damaged. Both such conduits and copper-

* One manufacturer of zippered sheath is Zippertubing Company of Los Angeles, California.

plated iron wire braids are commercially available. The most costly H-field shield is commercial galvanized electrical conduit in a completely welded system.

For temporary enclosures, galvanized iron or brass sheets may be used as shielding materials. The seams for such enclosures may be clamped or soldered. Soldering is not used in large enclosures because of expansion and corrosion problems. Copper foil tape with a conductive adhesive can be used as a finger-stock contact surface and for patching small holes, making bond straps, closing seams, etc. Finger stock for the doors can be "sticky fingers", a material with adhesive which requires no soldering.

If the breadboard for trying out the hardness design is to be a satisfactory model, certain construction details must be the same in the breadboard as in the prototype. They are:

- Cable braids must terminate on the outside of the enclosure skin.
- Braids must have both similar braid angles and density per unit area.
- Bond-strap impedances must be similar. This means that they should be dimensionally similar and bonded the same way.
- Enclosure dimensions must be similar.
- The grounding node scheme must be similar, and the cablings are routed in the same manner.

The harness in the breadboard will be much like that in the prototype, if the guard-shield systems are well-balanced and properly terminated. The exact arrangement of the wires in a bundle is not critical, nor is the precise routing, provided the rules of good design practices, as discussed in Chapter 6, are followed.

The breadboard system is then exposed to the field from the simulator, and the shielding effectiveness and adequacy are determined by comparing the appropriate probe-point responses, measured before and after the temporary hardening. When a pulse type simulator is used, responses in the form of peak voltages and ringing times are used for comparison.

From the viewpoint of cost effectiveness, it is then necessary to show that each change used in the breadboard is required. This is done by removing the breadboarded elements one at a time and noting the effect on hardness for each change.

After the breadboard shielding design is determined to be both necessary and sufficient, system response needs to be analyzed following the methods discussed earlier. The need for additional protective devices should also be examined at this point. Considerations for replacing any component by others which can withstand the impact of EMP should also be given at this time. However, any component or subsystem replacement should be kept to a minimum because of the obvious cost implication.

7.4.2 Hardness Verification Following Breadboard Design

Once the analysis indicates that a satisfactory hardening is achieved by the temporary hardening measures, the system or the relevant parts of the system should again be exposed to the simulator-generated fields, and responses should be measured. The measured data should then be analyzed again, taking care of any extrapolation required for the simulator fields. If this analysis shows that the required degree of hardness is achieved, appropriate permanent modification kits should be designed and fabricated following the breadboard design and construction closely. As a final checkout, it is necessary to expose the system or parts of the system again to the simulator fields and observe the response with the hardening modifications (according to modification kits) in place. If the results and associated analyses indicate that the required degree of hardening measures has been achieved, the system hardening should be certified accordingly.

While providing retrofit hardness to a system, it should be remembered that a great deal of hardness against EMP threat can be achieved without significant hardware changes. Many systems, for example, are vulnerable only for selected modes or conditions. If the system can be held in another mode for extended periods, or cycled quickly through the vulnerable mode, then an effective hardness will be realized without any significant expense for hardware changes. Another similar approach can be used to bypass system upset. In this approach, EMP is detected, and the system is recycled to restore the information lost during EMP period.

7.5 References

1. *Engineer's Handbook for Configuration and Technical Data Management*, Fort Belvoir, Virginia, December 1967.

2. Brown, G. L., Keith, R. M., and Shkabara, P., *Testing and Evaluation of Shielded Cables*, ANC 30R-27, American Nucleonics Corporation, Glendale, California, April 1970.

3. Brown, G. L., Ghose, R. N., Green, J. X., Putzer, E. J., Schwartz, K., *Electromagnetic Energy Coupling Studies - Design Practices*, ANC 30R-7, American Nucleonics Corporation, Glendale, California, February 1967.

4. Green, J. X., and Brown, G. L., *Hardness Assurance Program*, ANC 30R-25, American Nucleonics Corporation, Glendale, California, April 1970.

5. Vance, E. F., *EMP Hardness of Systems*, Fourth Symp. and Tech. Exhibition on Electromagnetic Compatibility, Zurich, March 1981.

CHAPTER 8

EMP Testing and Evaluation

Even when adequate tools and technology are available for vulnerability prediction by analysis, a validation of such prediction by tests is necessary to raise the confidence level. Although EMP is of nuclear burst origin, a validation test in an actual nuclear environment is neither practical nor necessary, and perhaps not even desirable. Tests in simulated EMP environments, therefore, constitute the only practical means of evaluating the system vulnerability or hardness. Again, since some extrapolations of test data have always to be made to make it correspond to the data which would result in a real-life nuclear EMP environment, one must carefully formulate test objectives, facilities and associated instrumentation to obtain meaningful test results. There was a significant development of various types of EMP simulation facilities, sensors, and instrumentations in the late sixties and early seventies. Merits and limitations of many different types of EMP facilities, sensors and instrumentations useful for EMP test validation, are discussed in this chapter.

Discussions on the design and implementation of tests which may be conducted during the acceptance or acquisition of a new system, and tests which may be carried out during the remainder of the system life cycle to assure its EMP hardness, are also included in this chapter.

8.1 Test Objectives

Although many analytical tools are available to predict system vulnerability or hardness, tests are necessary to ensure that the system is hard to the final "use environment". Since it is neither practical nor feasible to conduct tests in a true nuclear environment, alternative forms of test environments, facilitated by simulation are almost essential in EMP tests. To make the simulation tests meaningful and cost-effective, however, considerable judgment has to be exercised while formulating the test objectives, designing the tests, and interpreting and extrapolating the test data which will eventually be used to estimate the survivability of the system for any specified EMP threat. For EMP vulnerability assessment, a combination of tests and analyses is essential, since neither alone could provide the needed information.

Before any test is performed, a thorough understanding of factors which affect system response due to EMP is necessary. A total test program should then evolve from this understanding. Also, in the case of EMP, it is unlikely that a single test, no matter how carefully planned, will yield test data sufficient to design and fabricate a system and certify its EMP hardness. Thus, in general, several types of tests are useful and often necessary in this regard. Important among such types are:

- electromagnetic coupling tests
- component or subsystem characterization tests
- electronic susceptibility tests
- quality assurance tests
- hardness assurance or validation tests

For each type, one needs to formulate test objectives, plans and procedures, and means for data analysis and interpretation. Often, different test facilities are involved to conduct different types of tests.

The principal objective of electromagnetic coupling tests, for example, is to verify analytical predictions of coupling of fields, voltages or currents due to EMP at specified points due to penetration through shields, apertures, etc., and due to other coupling means such as antennas, cables, and power and utility lines. Similarly, the objective of the component characterization tests is to determine component parameters, and to assess the suitability of such components in a system which must perform its intended mission and be "hard" to the specified EMP threat. The objective of the electronic susceptibility tests is to experimentally verify the upset and burn-out characteristics of semiconducting devices and other electronic components. The objective of the quality assurance tests, as they relate to EMP, is similar to those involving the verification of uniform quality of units in production hardwares, excepting that the hardwares involved in this case are EMP hardness modification kits and EMP protection devices. Finally, the objective of the hardness validation tests is to verify by tests or experiments that the intended system mission will be achieved, notwithstanding the anticipated EMP threats. Once the test objectives are defined, actual tests to be performed in each case can be designed accordingly. The types and means of data collection, data interpretation and analysis leading to test objectives should be an integral part of the test design.

As the test objectives are somewhat unusual for the EMP threat, so are test facilities, and sensors and instrumentations used in such facilities. EMP test facilities are unusual because often such facilities have to simulate EMP which is characterrized by a very large electromagnetic field with an unusually fast rise-time. Characteristic requirements for EMP facilities, sensors and instrumentations, and various existing facilities meeting these requirements, are discussed in the following sections.

8.2 EMP Facilities

Different types of EMP tests, as discussed in previous sections, usually require different facilities. However, the most common EMP facility where, at least, the electromagnetic coupling tests, electronic susceptibility tests and hardness validation tests can be conducted, is the one which contains an EMP simulator.

The EMP simulator refers to a test tool or system designed to produce a known electromagnetic field which can be used to illuminate a system under test the same way as EMP could. As noted in Chapter 2, a typical EMP field can be "approximated" by the difference of two exponents. The waveform of such a field is shown in Figure 8.1, and the corresponding spectrum is shown in Figure 8.2. Although the vulnerability analysis should be based on actual waveform of EMP, the waveform and spectrum shown in Figures 8.1 and 8.2 are good test guidelines.

An idealized EMP facility should produce a waveshape similar to that shown in Figure 8.1, although there should be some overtest capability (with a higher magnitude of E and H fields by a factor of 2). Lower level generators can be used to investigate the effects of amplitude on arcing with appropriate extrapolations. The facility should produce arbitrary polarizations or have the capability of orienting the test object to account for polarization and direction of arrival. Also, an idealized EMP simulator should produce plane waves with a ratio of E/H equal to 120 π. In addition, the fields produced by such a facility should not be unduly affected by the test object. In practice, however, there is no facility which achieves all the desirable characteristics. In fact, adequate test data which can be extrapolated for the vulnerability analysis purpose can be obtained with simplified test facilities or simulators where one or more of the idealized criteria are missing. If, for example, the simulator pulse waveform is not the same as the worst-case threat waveform, or if it is not even similar to the test guideline waveshape as shown in Figure 8.1, one can obtain the test data for whatever simulator pulse waveform is available, and through a Fourier analysis determine the corresponding data for the waveform of concern. Similarly, even when the simulator produces only a cw signal, one can obtain an approximate spectrum for the test data by conducting several cw tests (with data points corresponding to the spectrum of interest shown in Figure 8.2). Moreover, when the wave-impedance for the simulated signal is not that of the free-space or $120\,\pi$, one can still extrapolate the test data to obtain the corresponding data which would have been obtained for the correct wave-impedance. Since as a final step toward EMP vulnerability of a system, some analysis has to be made, and since analytical tools are usually available to extrapolate test data obtained from a non-idealized facility to that which corresponds to the actual EMP threat, some flexibility exists in the simulator design.

Based on the flexibilities mentioned above, a simulator can be driven by either a pulse or a cw generator. It may be designed to simulate the plane wave of EMP originating from a distance, or a high-altitude burst, or it may be designed for close-in ground bursts, having a non-plane wave of peculiar wave impedance.

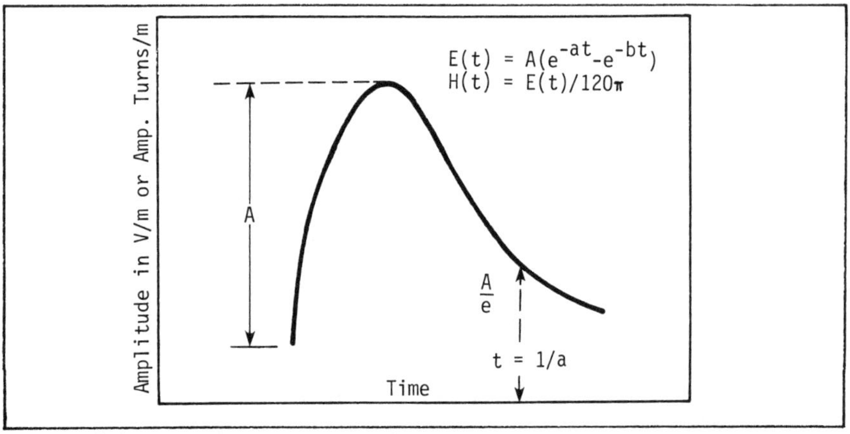

Figure 8.1 - Waveform Of Typical EMP.

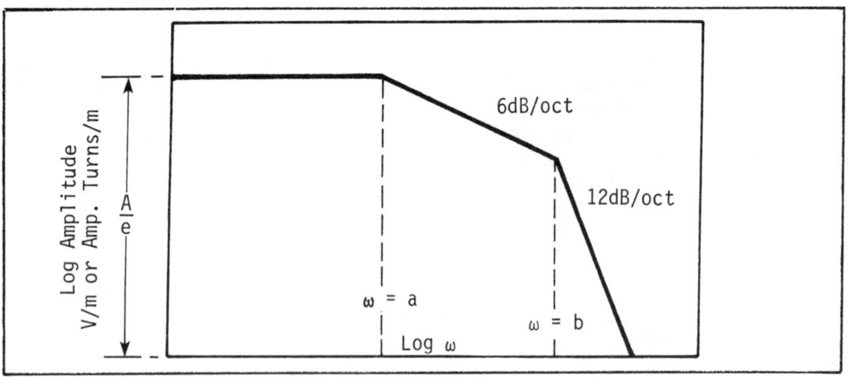

Figure 8.2 - Spectrum Of Typical EMP.

There are two basic types of EMP simulators. They are:

- the bounded wave simulator
- the radiating simulator

The distinction between the two types is based on whether the simulator attempts to contain the waves between two conductors or whether the waves are launched into free-space as radiated waves. The bounded-wave simulator generally produces a higher quality, more uniform field. Such a simulator, however, is relatively more expensive to build. The radiating EMP simulators are, perhaps, more versatile and can create more realistic propagating electromagnetic fields. But such simulators are undesirable sources of interference to nearby electronic systems not connected with the system under test.

8.2.1 Bounded Wave Simulator

This type of simulator produces electromagnetic fields either between two parallel plates or between a wire array, approximating two parallel plates. Basically, this simulator is a transmission line with the usable test volume in the middle between the plates or conductors. The field at the dominant transverse electromagnetic mode (TEM) for such transmission lines is indistinguishable from that in free-space propagation. The ratio of the magnitudes of the electric and magnetic fields is a function of the transmission line dimensions, and the ratio can be made equal to the free-space impedance. In both parallel-plate and wire-array type simulators, tapered sections form transition from relatively small dimensions of conventional coaxial cables or pulse generators up to a room-size or even bigger test-volume, and back down to a resistive load termination. The dimensions of such a simulator structure are determined by the size of objects to be tested and the desired frequency response. The electric field (E) is directed perpendicular to the plates but fringes at the edges. The magnetic field (H) is parallel to the plates at least in the uniform field region. Making use of conformal transformation, Baum* has analyzed the characteristics of fringe fields, field non-uniformity and impedance of transmission line simulators as a function of width of, and separation distance between, plates. This analysis provides a valuable insight into parallel-plate type EMP simulators desirable for their design and proper use. Constructions and characteristics of some existing bounded-wave simulators are further discussed later in this section.

---

* C. E. Baum, "Impedances and Field Distributions for Parallel Plate Transmission Line Simulators", EMP Sensor and Simulation Notes, Vol. I, No. 21, June 1966.

8.2.2 The Radiating Simulator

This type of simulator produces a radiating field launched by an antenna. Both horizontal and vertical polarizations for the radiating field can be achieved by selecting the appropriate antenna. The test objects, such as systems or subsystems, in such simulators are exposed to the simulator radiated field, and at responses at specific test points are measured. Both cw and pulse-type radiating signals can be excited by radiating simulators. Like the actual EMP field, the field radiated by this type of simulator is not contained between two conductors or plates. Although it is desirable to make the waveform of a radiated field similar to EMP waveform of concern, or at least similar to the test guideline waveform as shown in Figure 8.1, such a restriction is not necessary. This is because one can obtain by analysis the response due to a field with a specified waveform when test data due to a simulated field with a different, but known, waveform is available.

Many existing EMP simulators in the U. S. are of the radiating type. They include, for example, long-wire antennas, vertically polarized dipole, RES-1 (which is similar to the long-wire antenna but is airborne via a helicopter), etc.

Usually the radiating simulators have the following limitations:

- Inadequate response at low frequencies
- Non-uniform field intensities which fall off with distance from the source
- Appreciable component of the undesired polarized energy
- Large energy radiated as a source of interference to surrounding facilities

In general, the radiating simulators use basic antennas, modified in some ways to maximize low-frequency output and to minimize reflections from the ends of the antenna. Thus, the resistively-loaded dipole contains series resistors at periodic intervals to absorb outgoing energy, injected at the center of the dipole, and thereby minimize the energy that can be reflected back from the ends. The antennas, however, become less efficient as a consequence.

Cross polarization, or mixed polarization, has been a serious drawback to static charging of long-wire antennas. The collapse of the large vertical E-field built up prior to firing could induce effects almost as large as the build up of horizontal E-fields after firing.

Biconical antennas are good for launching high-frequency waves. The basic conic section above a ground plate, however, is often modified with resistive loading to minimize resonance due to finite length of the antenna. Also the top of the cone is deliberately flared out to increase the capacitance from the top of the antenna to ground. This improves the low-frequency response.

In one type of horizontal dipole EMP simulator, the antenna is basically a fat horizontal electric dipole, made fat by circular rings of wire. The ends of dipole are tied to the ground via conical transmission line. This enhances the low-frequency radiation. The antenna radiates, therefore, partly as an electric dipole of considerable length and partly as a vertical loop antenna of a large area.

8.2.3 ALECS Facility

One of the existing parallel plate transmission line type simulators is known as ALECS. An electrical equivalent of this simulator and a photograph of a scale model of B-1 Bomber undergoing EMP tests in this simulator facility are shown in Figures 8.3a and b. The signal generator for the simulator is an equivalent capacitor of about 1300 μμF, feeding a 95 ohm-transmission line terminated in its characteristic impedance. The rise-time (approximately 6 nanoseconds at 10 volts/meter) is determined primarily by the switch and the distributed inductance and load resistance. The decay time is determined by the circuit resistor-capacity-time constant (approximately 130 nanoseconds). Except for the reflections within the generator and the back radiation with its corresponding reflections due to unbalance-to-balance feed transition, the above factors define the shape of the simulated pulse in the reflection- and

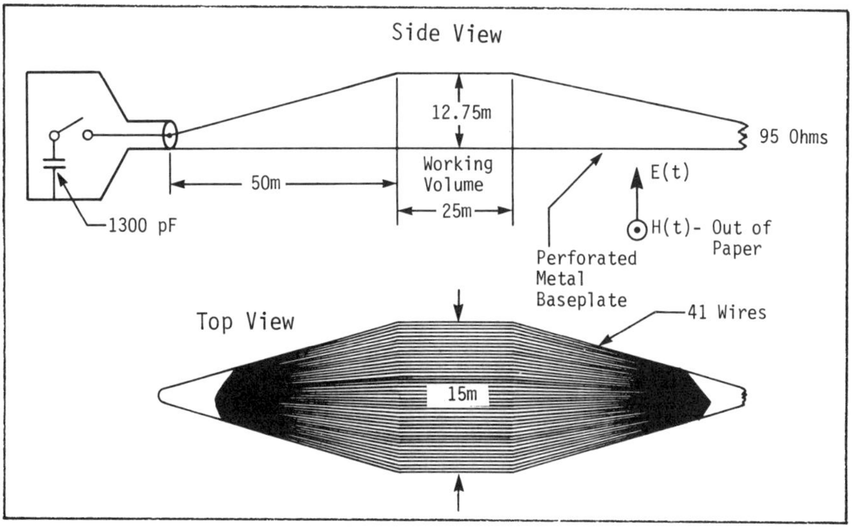

Figure 8.3a - Electrical Equivalent Of The ALECS EMP Simulator.

diffraction-free area of the working volume. An approximate variation of electric field (vertical) as a function time, typical of ALECS facility, is shown in Figure 8.4. The measured magnetic field (H) for this facility is nearly identical in shape as the E-field shown in Figure 8.4, the ratio of E/H being 120 at least within the working volume. The working volume for the facility is 26.9 x 15.24 x 12.75 cubic meters. The shape of the spectrum of the vertical electric field, as obtained from the Fourier transform is shown in Figure 8.5.

The magnitude of the electric field in this facility is approximately the capacitor charge-voltage divided by the spacing between the plates. The actual electric field is about 70 to 80 percent of this calculated value. The generator for this facility uses a Van de Graff charging belt technique and can produce a voltage on the order of 2.3×10^6 volts reliably.

Figure 8.3b - A photograph of the ALECS Facility with a B-1 Bomber Scale Model Undergoing EMP Tests.

A pressurized gap switch is used to control and hold the capacitor charge voltage. The gap is pressurized with a variety of gases including dry nitrogen and sulphur hexafloride. The gap is controlled by a mechanical plunger which reduces the gap spacing in about a millisecond, resulting in a gas breakdown and discharge of the capacitor into the line. The minimum field produced by this generator in the working volume is about 3KV/meter. Lower level repetitive pulsers of the charged capacitive type are easily adapted to this facility for low-level studies. The load for the transmission line is salt water.

The ALECS facility produces only one polarization, with E field normal to the parallel plates. Equivalent polarization changes, however, can be achieved by changing the orientation of the test object. Also, rotation of test objects can simulate a change in the wave direction of arrival.

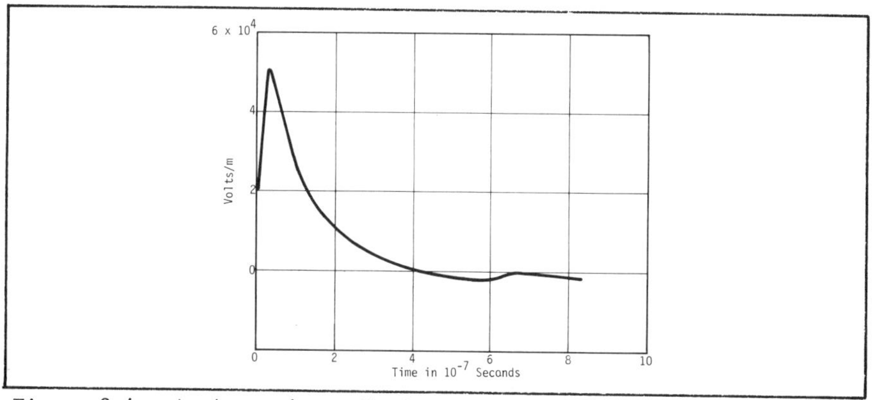

Figure 8.4 - An Approximate Variation Of Vertical Electric Field With Time For ALECS Facility.

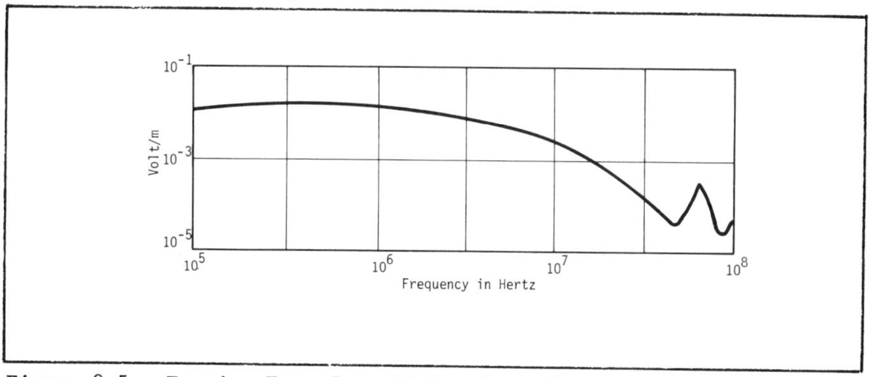

Figure 8.5 - Fourier Transform Of Vertical Electric Field Shown In Figure 8.4.

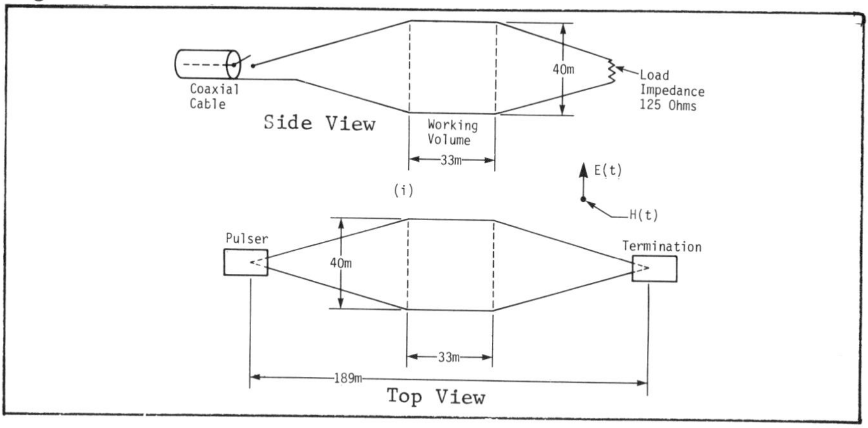

Figure 8.6a - Schematic Arrangement Of ARES EMP Simulation Facility.

8.2.4 ARES Facility

Another existing transmission-line type EMP simulator, which is larger (189 meters long and 40 meters high) than ALECS facility, is ARES (Advanced Research EMP Simulator). The schematic arrangement for this facility and its photograph are shown in Figure 8.6a and b.

The generator is an open-ended transmission line whose half-wavelength resonance occurs at 17 MHz. The rise-time for this facility is a function of the inductance of the switch and the impedance of the facility line. The general decay of the pulse is controlled by the capacitance of the transmission-line generator and the impedance of the facility line. The decay is actually a step-by-step drop due to reflections in the generator, as shown in Figure 8.7.

An approximate shape of the magnetic field waveform is shown in Figure 8.8. An approximate spectrum of this field is shown in Figure 8.9.

Figure 8.6b - Photograph Of The ARES Facility.

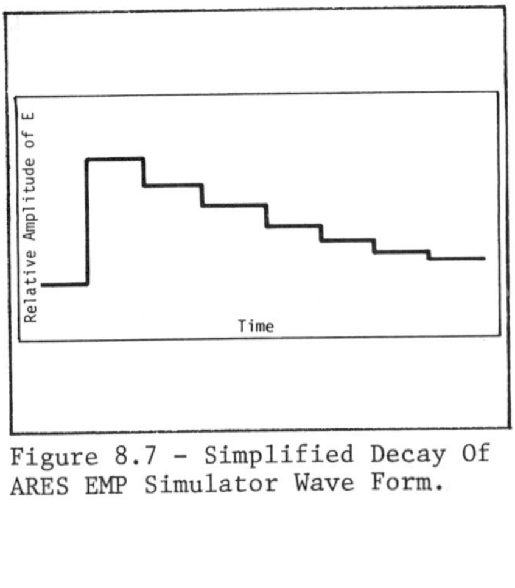

Figure 8.7 - Simplified Decay Of ARES EMP Simulator Wave Form.

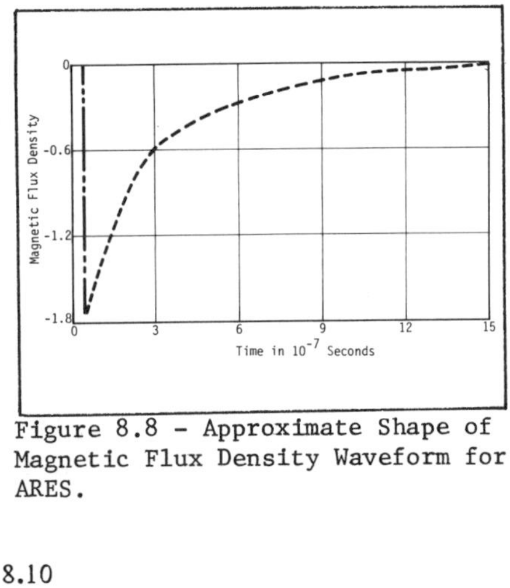

Figure 8.8 - Approximate Shape of Magnetic Flux Density Waveform for ARES.

The magnitude of the electric field pulse in the ARES facility is the charging voltage divided by the plate separation distance d. Since the ARES plate separation is much greater than that of the ALECS facility, a larger charge voltage is required to produce the same vertical E-field in the working volume. The working volume for ARES is 40 x 33 x 40 cubic meters. The maximum field in the working volume is 100 KV/meter, and the minimum field is about 15 KV/meter. Repetitive pulse generators for low-level testing, with charging voltages of 5 and 50 KV, are available in the facility.

The generator for the facility is best represented by a transmission line as shown in Figure 8.10. The energy for ARES is deposited in the capacitance of the line using the Van de Graff charging belt techniques. The impedance of the generator coaxial line is 15 ohms, and the load or line impedance is about 125 ohms. The line is open-ended and feeds a large mismatch. The generator energy is discharged through a multi-element gap switch pressured by gases, similar to those used in ALECS facility. The rise-time of the field is controlled by the switch characteristics, and is variable with the field amplitude level. For example, the rise-time is shorter and more stable at high level, but is, in general, between 5 and 10 nanoseconds for all levels. The generator may be discharged by using a mechanical pin as in ALECS, or by an electronic trigger.

The electromagnetic characteristics of a bounded-wave vertically-polarized simulator such as ARES are well understood. The geometry and dimensions of the transmission line determine its impedance. For the ARES geometry, the height-to-width ratio is equal to 1.0; however, the ground is effectively an infinite conducting plane, so the equivalent height-to-width ratio for a two-plate stripline in free-space is 2.0. This implies an impedance of approximately 127 ohms.

Another mammoth simulator, known as the Trestle Simulator, is designed to have two pulsers, each producing a pulse of about 5 megavolts, and providing a horizontal electric field. A 61-meters-by-61-meters wooden

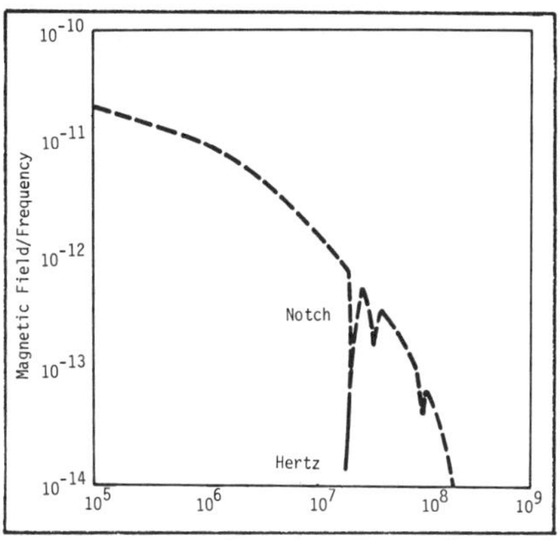

Figure 8.9 - Fourier Transform of Magentic Flux Density Waveform in Fig. 8.8

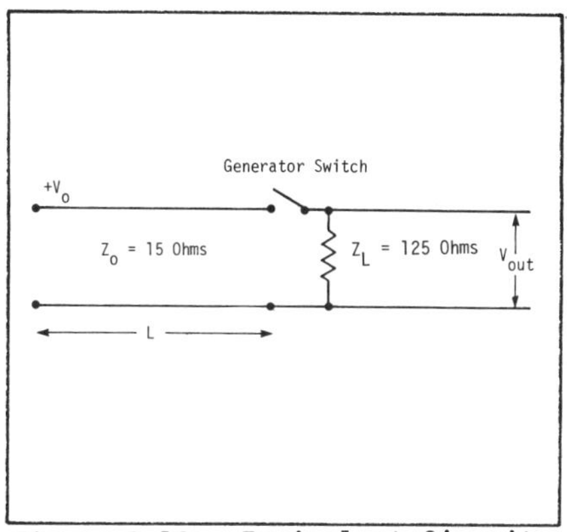

Figure 8.10 - Equivalent Circuit of Generator and Load of ARES Simulator.

platform in this facility may support a test aircraft 36 meters above the ground between two vertical wire planes which form the two-plate transmission line.

In addition, there exists a number of simulators, known as HPD (Horizontally Polarized Dipole) and VPD (Vertically Polarized Dipole), designed by Air Force Weapons Laboratory and the TEMPS (Transportable EMP Simulator) at Harry Diamond Laboratories of the United States Army. A summary of these facilities is provided by Giles*.

8.2.5 Small EMP Test Facility

EMP simulators such as ALECS and ARES have large working volumes such that large test objects, such as airplanes, can be tested directly. Often, the required test volume is not large, and an in-house EMP test facility may be desired for many electronic and weapon system manufacturers and users. Such a facility may be built relatively inexpensively and, notwithstanding some limitations, could be convenient and cost-effective.

The plan view and side view of such a facility** are shown in Figure 8.11. As seen from the figure, the parallel plates are 1.8 meters wide and 4.8 meters long. The plates are tapered at both ends to provide convenient points to connect the source generator and the load. The separation distance between the plates is approximately 0.75 meters. The equipment or object to be tested is placed in the control region of the parallel-plate transmission line.

The parallel plates can be made of wire meshes stretched over wooden frames. The transmission line is excited by a high-voltage dc source through the discharge of a capacitor, as shown in Figure 8.12. A spark-gap switch, which operates manually to activate a gas valve and permit the pressurized air in the spark gap to escape, is used during the discharge. The maximum dc voltage across the transmission line can be on the order of 50 KV. The facility, therefore, can have the capacity to generate an electric field of approximately 66.67 KV/meter at best near the center of the transmission line. A variable electric field is obtained by controlling the primary voltage of the high-voltage transformer.

The parallel-plate transmission line shown in Figure 8.11 does not have the desired degree of field uniformity throughout the available test volume. In

---

* J. C. Giles, "Simulating the Nuclear Electromagnetic Pulse," Military Electronics/Countermeasures, August 1977.

** The geometry and construction of a similar small EMP test facility are outlined in "NEMP Simulation and Tests in Switzerland", T. Rudy, J. Bertuchez, and B. Wamister, Third Symp. and Tech. Exhibition on EMC, Rotterdam, May 1979.

particular, the fringe field at the edges may be far from that required for the tests. However, if the test volume is kept small, such that it does not occupy more than one-third of the width of the transmission line and more than one-half of the distance of separation, meaningful tests can be conducted with such a facility.

Prior to testing, it is necessary to measure electric and magnetic fields within the test volume to calibrate the facility. The electric and magnetic field intensities at any point within the test volume can be measured by inserting a probe or a loop within the test volume. The probe needs to be oriented vertically, along the direction of the electric field, and a calibrated voltage divider can be used to reduce the voltage to be measured by a known ratio. The reduced voltage can then be measured by a cathode ray oscilloscope. One arrangement for introducing a vertical electric field sensing probe

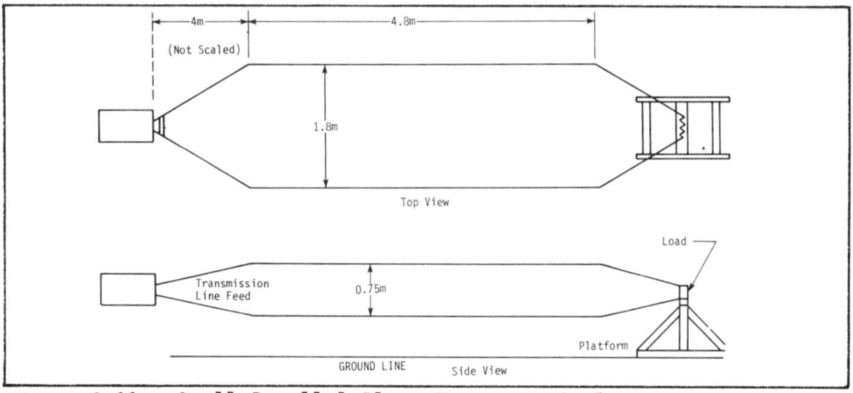

Figure 8.11 - Small Parallel-Plate Type EMP Simulator.

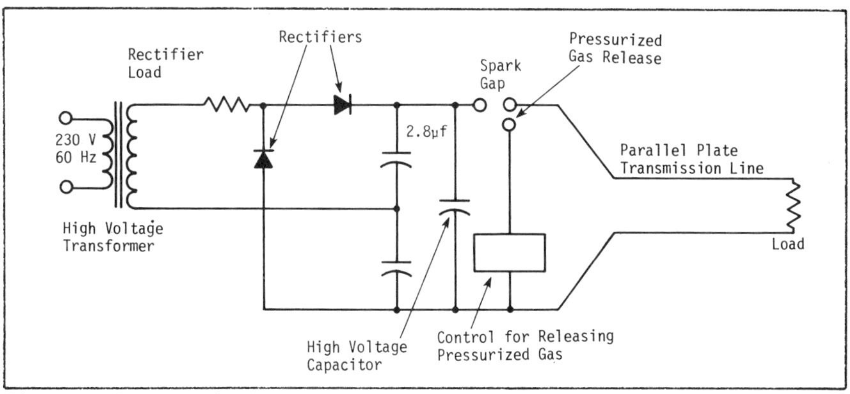

Figure 8.12 - Equivalent Circuit of the Generator for Simulator shown in Fig. 8.11.

with a voltage divider, so that the field is least disturbed by the probe and its support structure, is shown in Figure 8.13. The support structures for the probe could be a glass bar about 3 cm wide and 1 cm thick and 15 cm long. The two parallel wires of the transmission line for the probe could be mounted at the edges of the glass bar. The probe and voltage divider could be mounted at one end of the bar while the other end is clamped to a vertical post. The vertical post could be provided with a scale so that one can readily determine the height of the center of the probe with respect to the ground plane or the bottom plate of the simulator. By raising or lowering the glass bar and by moving the vertical post horizontally along the transmission line, one can measure the vertical field intensity at most points of concern within the test volume.

To measure the magnetic field intensity, the probe should be replaced by a loop, and the same movable support structure could be used for the loop, the plane of which will be parallel to the vertical direction of the simulator.

The voltage divider to be used may reduce the voltage by a ratio of 100:1, so that the actual voltage to be measured by the oscilloscope will be on the order of 30-35 volts.

The probe and the loop should be calibrated by feeding a test signal of known voltage amplitude across the simulator transmission line plates, and noting the ratio of the voltage measured by the calibrated scope to that of the signal impressed across the simulator plates. The ratio should be measured at various frequencies of interest, and an attempt should be made to keep this ratio about uniform in the frequency range of 100 KHz to 20 MHz.

In the test facility shown in Figure 8.11, the simulated EMP may have a rise-time ranging from 15 to 50 nanoseconds and a pulsewidth of approximately 1/2 - 1 microsecond. There will be an exponential decay of the pulse after it has reached its peak value. In most cases of interest, no provision

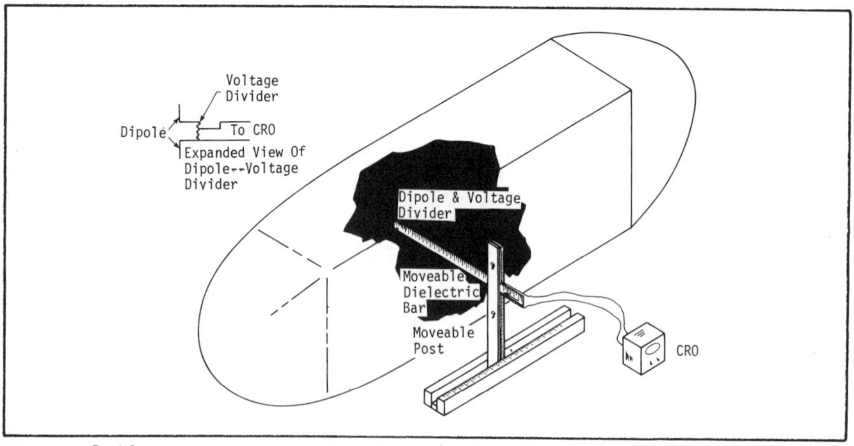

Figure 8.13 - Arrangement For Introducing Electric Field Sensor in the Simulator Shown In Fig. 8.11.

needs to be made to repeat the simulated EMP in quick succession. Each test, however, should be repeated at least four times to ensure the repeatability of the result.

In many electronic systems, the susceptibility due to EMP may vary depending on the polarization of EMP field. This is primarily because the coupling factors for various coupling mechanisms, such as apertures, seams, etc., change with the orientation of the test object with respect to the direction of the impressed EMP field vector. To ensure that EMP vulnerability of the equipment for the worst-case environment is determined, it is necessary to change the polarization of the simulated EMP field. Since this is not practical in the simplified test facility of Figure 8.11, test objects can be rotated or appropriately oriented within the test volume.

8.2.6 Long-Wire Antenna Simulators

One existing long-wire antenna simulator was built by Sandia Laboratories and is located at Albuquerque, New Mexico. It is a low-level EMP simulator facility which produces an E-field of about 400 volts/meter along the direction of the antenna at about 15 meters on the centerline and 1.5 meters off the ground. The schematic arrangement for this facility is shown in Figure 8.14. It is a physically tapered line which is about 0.3 meters in the center and 0.1 meter at the end. The larger diameter in the center results in a greater field. The peak magnitude of the field is determined by the center section of the antenna. The antenna is also electrically tapered at about 6-meter intervals by carbon resistors as shown in the figure. There are 24 resistors on each side varying from 6 ohms to 250 ohms. The antenna is charged by 66.7 K ohms/meter charging lines. These lines are made of lossy materials to prevent radiation. They are located 30 meters from the center of the antenna. Two 25 K volts generators, connected in series with an earth return, are used to charge the antenna. The dry nitrogen pressurized switch is

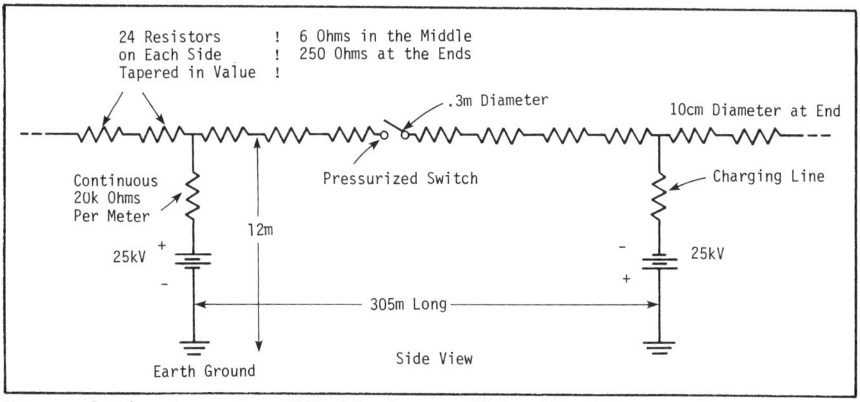

Figure 8.14 - Longwire EMP Simulator Of Sandia Laboratories.

set at about 35 ,and breaks down automatically at about 30 KV, producing 20 pulse per seconds. The rise time of the switching action is about 2.5 nanoseconds. The electric field at 15 meters out on the ground plane is shown in Figure 8.15.

8.2.7 Vertical Dipole Simulator

Another existing radiating antenna type simulator is a vertical dipole simulator. This facility is basically a resistively-loaded conical antenna over a ground plane. The schematic arrangement for the antenna is shown in Figure 8.16. The cone is about 27 meters high and 39 meters in diameter at the base. The antenna is driven by a 1.75 M volts pulse generator and can produce peak electric fields on the order of 10-15 KV/meter at about 100 meters from the apex of the cone. An approximate waveform of the pulse generated by the vertical dipole simulator is shown in Figure 8.17.

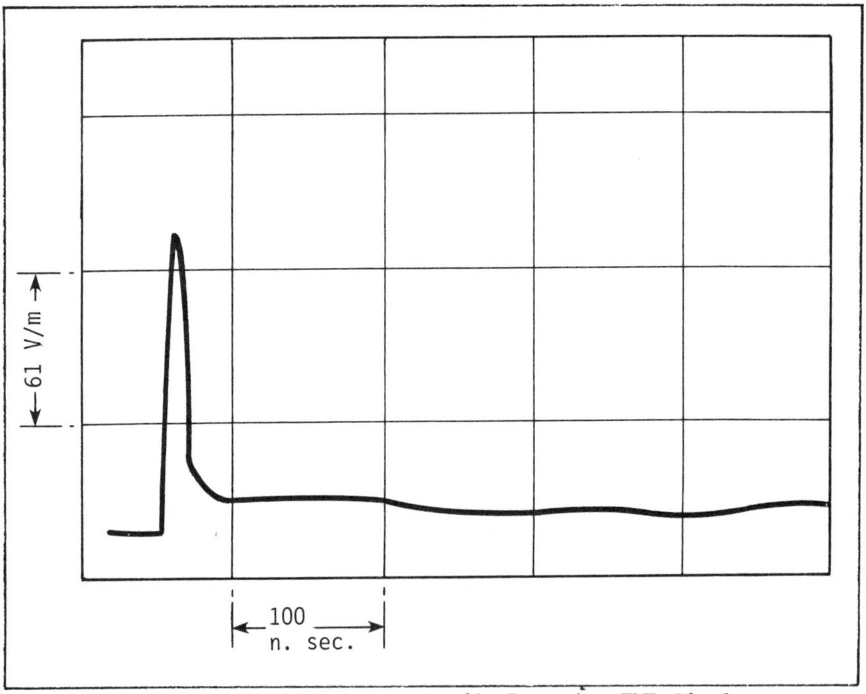

Figure 8.15 - Electric Field From Sandia Longwire EMP Simulator At 15m out and 2.5m Above Earth.

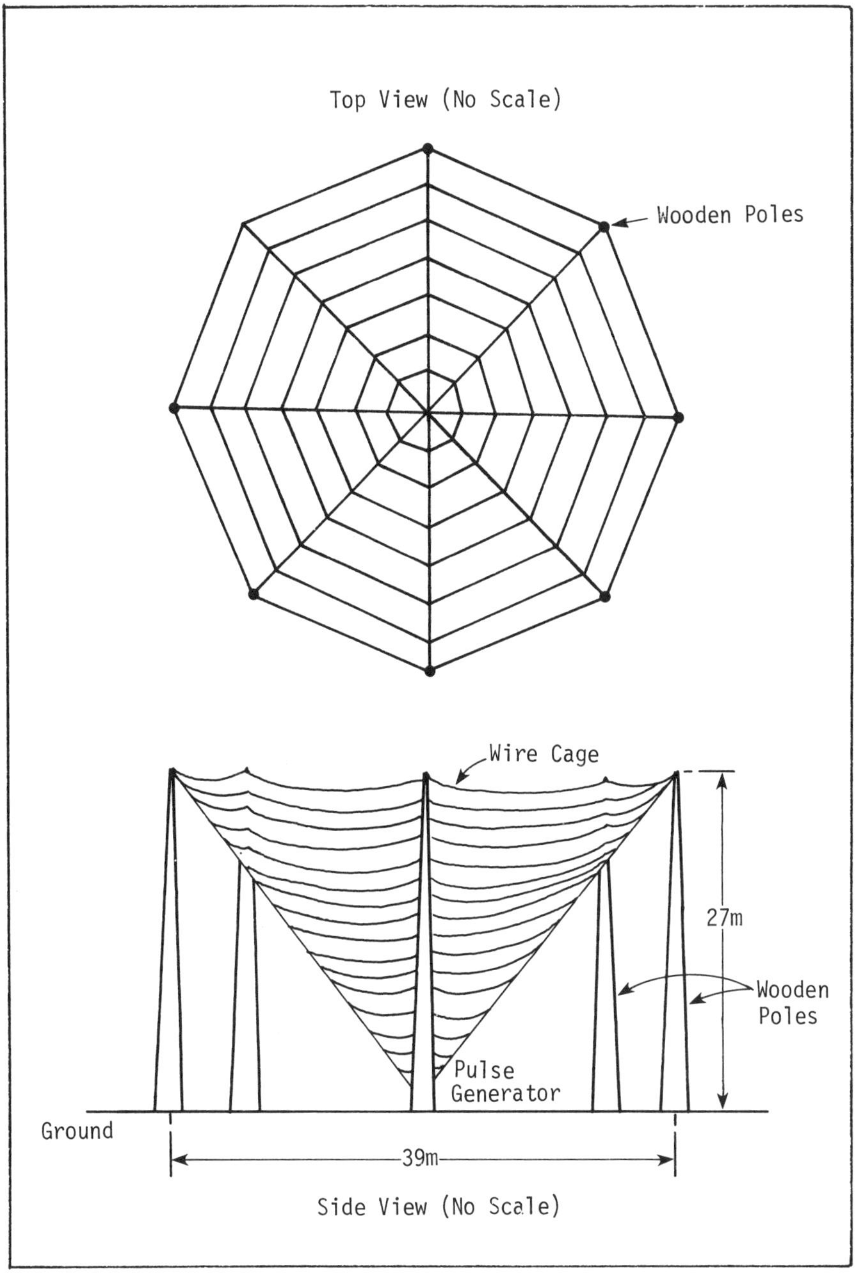

Figure 8.16 - Vertical Dipole EMP Simulator.

8.2.8 Cable Driver

As noted in Chapter 3, a principal mode of EMP energy coupling is the generation of currents in various cables of electronic or weapon systems. These currents can be large enough to cause damages to circuit components or to result in system upset. To design a system with an objective to avoid system vulnerability due to EMP-induced cable coupling, and to predict vulnerability of an existing system due to such coupling, one needs to estimate:

a. sheath currents in a cable.

b. currents and voltages in the individual conductors of both shielded and unshielded multiconductor cables.

c. currents in two-wire transmission line, particularly near the earth.

d. currents generated in large loops where the propagation time across the loop is large compared to the rise-time of EMP field.

Some of these parameters could be easily determined experimentally with a relatively small EMP simulator exclusively designed for cables. Such simulators are often referred to as cable drivers. The schematic arrangement and photographs of a typical cable driver are shown in Figures 8.18a, b and c. The capacitor is charged by a high-voltage power supply. The cable driver is driven by the current that flows when the capacitor is discharged by the air-gap. Usually the cable driver is used to test shielding effectiveness of a cable, measure its bulk current due to simulated pulse, and for other diagnostic tests.

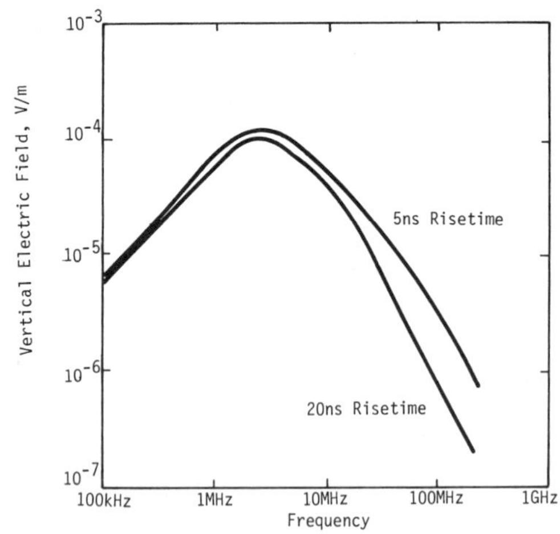

Figure 8.17 - Approximate Spectral Characteristics Of The Field Generated By Vertical Dipole Simulator.

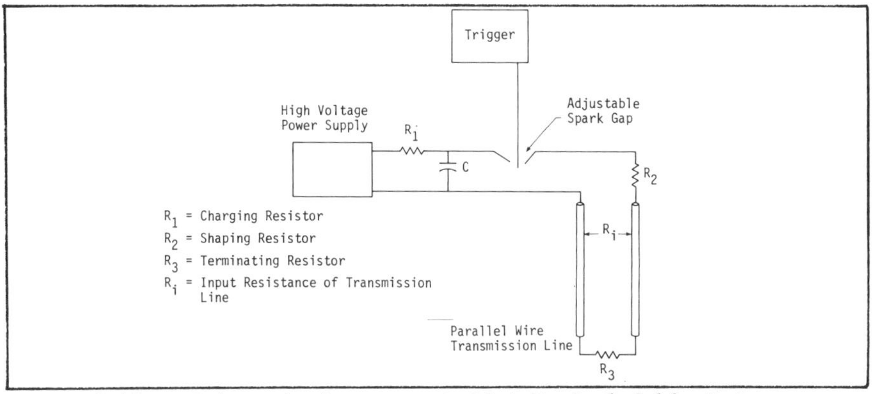

Figure 8.18a - Schematic Arrangement Of A Typical Cable Driver.

Figure 8.18b - CAble Driver Test Set-up For Long Cables.

Figure 8.18c - Adjustable Spark-gap.

8.2.9 Continuous Wave Tests

Simulators and test facilities discussed so far generate a pulse-like waveform for the simulated field or voltage or current. EMP coupling tests, however, can also be conducted with cw test signals. In fact, recent EMP coupling tests have gone more and more toward cw measurements, particularly now that fully-automated transfer function measurements are possible using instrumentation equipment.

In cw measurements, test data is obtained by setting a transmitter to a fixed frequency (pure sine wave) and measuring the steady-state input field or current in a system, or the amplitude and phase responses of a system, or both. Dividing the output by the input yields a transfer function value at the selected frequency. Repeating this process for many frequencies and extrapolating justifiably for others, one obtains the transfer function for the system which can then be made amenable to Fourier analysis.

The process of setting transmitter frequencies, measuring amplitudes and phases, and recording the data as shown schematically in Figure 8.17, have been automated in modern test systems. Often such machines include such features as programmable frequency selection, automatic frequency-stepping and measurements, automatic noise measurements, rough plots in real time, and automatic data logging on a magnetic tape. Such an automation leaves human action during testing limited to the setting of probes and sensors at appropriate places, annotating the tapes and other recorded data and general monitoring. The importance of a rough-transfer function plot available in real time cannot be overemphasized. It provides the only clue to how well the test is proceeding.

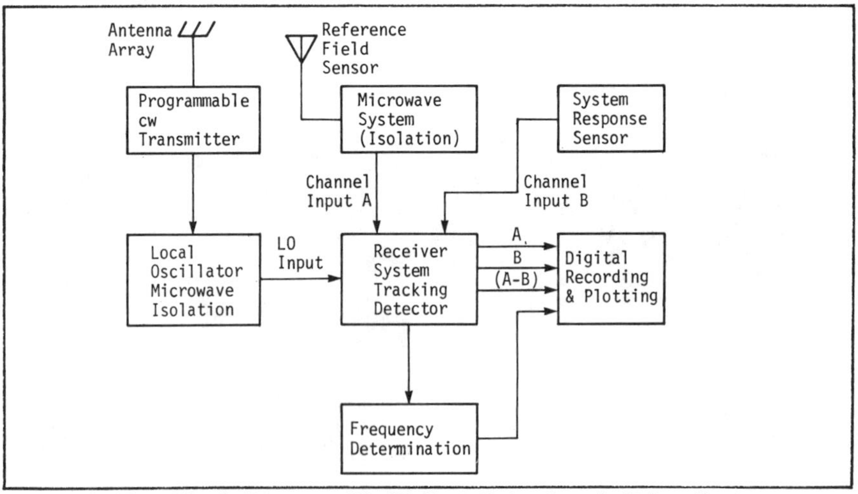

Figure 8.19 - Block Diagram Of CW Test Data Acquisition System.

Cw measurements can usually be obtained with an accuracy of about a dB in amplitude over the entire frequency range of interest, and to within a few degrees for the phase measurements. Test systems have been built to cover frequencies from 10 KHz to 60 MHz. Such systems are almost unlimited in dynamic range since they can be made self-adjusting. The receiver used in such systems is one of very narrow bandwidth (a few hertz), and its sensitivity is very high. Out-of-band noise rejection is also excellent due to the narrow bandwidth, often the principal limiting noise source being the internal noise of the receiver. Sometimes stray noise pick-up, due to improper shielding of instrumentation can also be a serious noise source during cw tests.

Cw measurement systems can clearly provide higher-quality EM coupling data, when properly used, but there are certain system characteristics and parameters that cannot be determined by cw measurements. For example, one of the advantages of cw measurements is that low-level signals can be used for measurements. However, because of this low-level, and because cw measurements do not represent an EMP waveform, some of the sources of problems such as exciting an arcing, or causing a breakdown of surge anestors, or causing an electronic component or circuit to be non-linear, etc., escape tests and evaluation. For this reason, it is necessary to complement cw tests with high-amplitude pulse tests for a complete vulnerability assessment.

8.2.10 Current Injection Tests

In addition to testing a system with simulated EMP or cw signals, one can obtain useful information on the characteristics and susceptibility of certain subsystems and cables by a current injection test technique. As discussed in Chapter 3, EMP field of any given waveform sets up current transients at the input of various subsystems or at the interface of subsystems. If the amplitudes and waveforms of such a current transient are known at a given input or interface of subsystems or cables, either through analysis or from past measurements of the transient, then one can assess experimentally the EMP susceptibility of the subsystem by injecting a simulated current transient at the input or interface under consideration and observing the response. Thus, it will not be necessary to subject the whole system to the specified EMP threat for the purposes of vulnerability assessment.

The current injection test often turns out to be a very practical and economical approach to test subsystem susceptibility, particularly when only one or a few such subsystems dominate the overall system vulnerability problem. Usually, the current injection tests are conducted in a laboratory, although field application of such tests may be required, at times, for hardness assurance.

There are various reasons why the current injection tests are usually recommended. First, an analysis and test of each individual circuit or system is prohibitive in time and cost. This suggests tests of selective circuits or subsystems based on their sensitivities. Secondly, one can test selective subsystems or a group of subsystems by one current injection. Moreover, since the current injection tests are usually conducted in a laboratory, it is not so expensive.

Where it is possible to do so, current injection tests should use accurate simulations of the expected waveform. When such data is not available, tests must be conducted on the basis of a parametric investigation to determine the effects of amplitude, frequency, decay rate, etc.

While conducting current injection tests, the simulated electromagnetic pulses are increased in amplitude, in steps, until a failure is induced. One may inject a single pulse, or pulses may be injected on a repetitive basis. When repetitive pulses are used, counting the number of pulses before a failure occurs provides a basis for statistical estimation of the probability of failure for various current levels.

To date, EMP pulse generators or function generators of various types have been developed for current injection tests. These generators can excite one subsystem or cable at a time or several of them at once. Frequency and decay rate are variable in most function generators. In some function generators the synthesis of the required waveform is accomplished by superimposition of several damped sine-wave generators where each may have a different amplitude, a different frequency, a different decay rate and a different start time or phase. During current injection tests, it is necessary to avoid stray pick-up of noise. Usually appropriate shielding of instrumentation avoids this problem.

In summary, current injection tests are a valuable aid to EMP testing. They constitute excellent tools for failure mode studies and provide an efficient method for screening large numbers of circuits for upset or permanent damage effects. Once the most vulnerable circuits or subsystems have been identified, individual circuit tests or analyses may be justified to provide the necessary detailed understanding of system failures.

The current injection tests have their limitations, however. One of them, for example, is that it is virtually impossible to place meaningful uncertainty limits on the test data, since the most serious errors are likely to result for unknown imperfections in the simulation. One can, in most cases, however, estimate the upper bounds of any error due to small simulation imperfections and use an appropriate correction factor for the test data, if needed.

8.3 Instrumentations and Sensors

Because of the unusual characteristics of EMP or simulated EMP, instrumentations and sensors suitable for EMP tests need to be designed and constructed accordingly. For example, since EMP is a short-term transient phenomenon, test equipment must be of the wideband type that can cover a frequency range of about 100 KHz to 100 MHz.

Although the characteristics of instrumentations and sensors for EMP tests are often unusual, most EMP measurements involve traditional electrical qualities such as the voltage or current or field intensity. Typical sensors used for EMP measurements, therefore, are:

- Current probes
- Voltage probes
- Field-strength measuring devices

8.3.1 Current Probes

Current probes are used to measure EMP-induced current flowing through a wire or a bundle of wires in a system. Usually such probes clamp over the wire or a bundle of wires during the measurement. A typical bulk current probe is shown in Figure 8.20. Basically, a current probe is a current transformer whose primary is the wire or wire bundle carrying the current, and secondary is a winding on a core which provides the output. Usually the device has a ferrite core with a winding spiralling about the core. For the clamp-on type current probe, air gaps are formed at the hinge and clamp. The core and winding have a metal case for quasi-electrostatic shielding. The static shield has a circumferential slot on the inside surface to allow flux penetration.

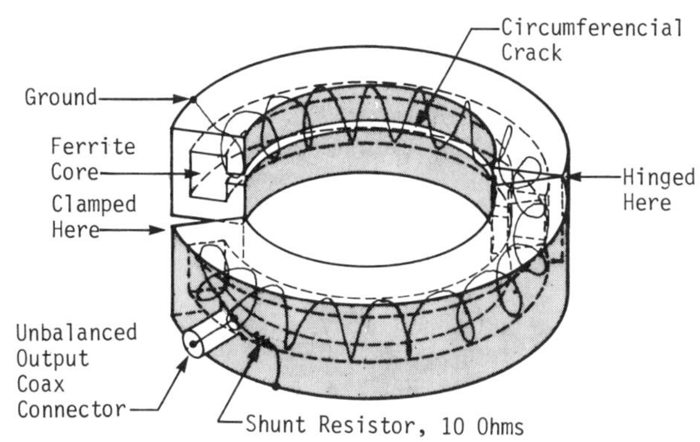

Figure 8.20 - A Typical Bulk Current Probe.

The principal characteristics of a current probe are its (a) transfer impedance measured in volts out per unit ampre through current, (b) the strong field pick-up, (c) the insertion impedance, and (d) the capacity of the probe to cable. The transfer impedance is generally quoted for a 50-ohm cable and 50 ohm termination. Most probes have a transfer impedance between 0.5 and 10 ohms in the flat frequency region. The insertion impedance for a 1 V/A probe in the flat frequency region is about 0.02 ohm, which is negligible for most applications.

Usually, problems in current measurement arise due to the capacitive coupling of the probe to the cable or wire whose current is being measured. This effect can be reduced by using continuous strings of ferrite beads on the probe cable. The beads both increase the impedance in series with the capacitance, and introduce loss to dampen currents on the cable.

Current probes should be calibrated prior to conducting tests. This is usually done at a low level, and the probe's transfer impedance is plotted as a function of frequency. If the probes are a new type and are going to be used for high-level testing, extremely high current pulses should be used to determine linearity or burnout levels. Currents as high as 12,000 amps (peak-to-peak) have been measured with some probes.

Variations of the basic current probe shown in Figure 8.20 include non-clamp-on type with no air gap. This type of probe can be custom-built to fit in tight spots, but does not have the bandwidth obtained by air-gap types. Small probes which may fit over a single wire have been developed with typical transfer impedance of 5.0 ohms over a frequency band from 1 KHz to 1 GHz.

Bulk current probes respond to the magnetic field. Hence, they will not measure currents beneath overall shields at frequencies above that which constitutes a skin depth.

8.3.2 Voltage Probes

For both diagnostic and system evaluation tests, it is often necessary to measure voltage-waveform at various points in a circuit or a system. Also, for certain kinds of permanent damage in film resistors and in integrated circuits, the failures seem to depend on voltage and nothing else. It is natural, therefore, to want to include voltage measurements as a part of any well-conceived EMP test program.

Since absolute potential cannot be defined, voltage for practical purposes is defined as the potential difference between a point of interest and a reference point, often termed as "ground". This reference is not strictly defined, except in relative terms.

A differential voltage probe is required for nearly all voltage measurements or electronic components, because the circuitry is floating relative to the ground. Principal errors in voltage measurements by a voltage probe could be due to:

- Loading effect
- Deterioration of common mode rejection
- Magnetic field pick-up

To illustrate the sources of error, one may consider a typical circuit where the voltage measurement should be made, and the circuit of a voltage probe as shown in Figure 8.21. To minimize error in voltage measurements, the input impedances Z_1, Z_2, and Z_3 of the voltage-probe circuit must be very large relative to the impedances looking back into the circuit of interest, that is, Z_1', Z_2' and Z_3'. If, however, the Z_1, Z_2, and Z_3 are not large compared to Z_1', Z_2', and Z_3', a loading error occurs, distorting effectively the voltage to be measured. Typical values of Z_1, Z_2, and Z_3 are 1 to 10 Megohms of resistance and 5 to 20 picofarad of capacitance.

Similarly, when Z_2' in the circuit is very small, and all other impedances in the circuit are large, a larger potential-to-ground relative to the voltage across Z_2' may exist. Input amplifier must cancel the effect of large potential and measure only the differential voltage. This cancelling action, often referred to as common mode rejection, deteriorates as the frequency is increased. A very large common mode signal may overdrive the semiconductor input device of the voltage probe resulting in a greater error in voltage measurement.

The error due to magnetic field pick-up occurs because the input leads of the voltage probe along with the impedances Z_2, and Z_2' form a loop. The undesired voltage induced in this loop is superimposed on the voltage being measured, and thereby introduces an error. In the time domain, the open-circuit voltage induced by the loop is $\mu A \frac{dH}{dt}$, μ, A, H and t being respectively,

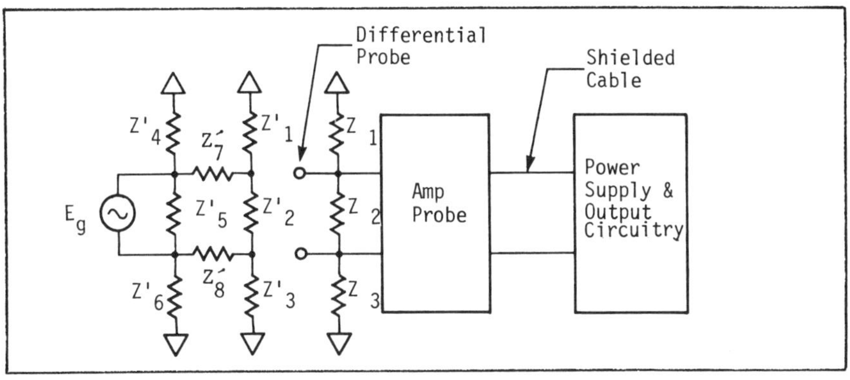

Figure 8.21 - Typical Measurement By A Voltage Probe.

the permeability (that of free-space in this case), loop area, magnetic field intensity and time. The corresponding open-circuit voltage induced in the loop for a sinusoidal H is $j\omega\mu AH$, where is the angular frequency. Thus the error due to magnetic field pickup increases for fast rates of change of H or at higher frequencies. Also, when proper impedance requirements, that is Z_2', Z_2 among others are met, almost all the induced voltage appears across the probe, thus aggravating the problem. It is evident that extending the leads from the probe tips increases the magnetic field pick-up problem. Another source of error associated with the voltage probe is improper or inadequate shielding of input and power leads.

In general, voltage measurements employing the voltage probe as discussed above are discouraged because of the difficulty in making such measurements accurately. More specifically, voltage measurements should be made with a differential probe as close to the load as possible. Unfortunately, it is often very difficult to gain access to the actual load in the electronic package in the field.

In using the voltage probe, it is desirable to make a quick check by reversing the input leads. The observed waveform should reverse in polarity unless there is some common-mode rejection problem. The error due to magnetic field pick-up, however, cannot be detected by this means.

8.3.3 Field Strength Measuring Devices

Typical sensor used for field-strength measuring devices are dipoles and loops. A short dipole measures the electric field intensity E, while a loop measures the time derivative of the magnetic field intensity, that is $\dot{H}$. In the system analysis, however, one uses the field parameters E and H (not $\dot{H}$). It is possible to measure H by shorting the output of a loop and measuring the resulting current. Similarly, if required, one can measure E by shorting the output of dipole and measuring the resulting current.

Parametric relations among field components and the sensor-measured quantities are as follows:

<u>Open Short Dipole</u>

$$V_{OC} = 2H_eE \simeq HE \quad \text{for } H \ll \lambda \tag{8.1}$$

where,

V_{OC} = open-circuit voltage at dipole terminals

H_e = effective height of the dipole (each half)

H = height of the dipole (each half)

λ = wavelength of the field to be measured

E = electric field along the dipole.

A linearly decreasing current distribution from the base (center of the dipole) to end tip is assumed to relate H in terms of H_e.

Open-Circuit Loop

$$V_{OC} = -\mu A\dot{H} \tag{8.2}$$

where,

V_{OC} = open-circuit voltage at loop terminals

A = area of the loop

μ = permeability

$\dot{H}$ = time derivative of the magnitude field intensity H

Short-Circuited Dipole

$$I_{SCD} = \frac{V_{oc}}{Z_a} \simeq HC\dot{E} \simeq i\omega HCE \tag{8.3}$$

where,

C = capacitance of the dipole

$\dot{E}$ = time derivative of E

ω = angular frequency of field to be measured

I_{SCD} = short-circuit current

Z_A = antenna impedance

The expression I_{SCD} for the short-circuit current as given in Equation 8-3 is based on relation

$$V_{OC} = HE$$

and,

$$Z_A = R_A + \frac{1}{i\omega C} \simeq \frac{1}{i\omega C} \tag{8.4}$$

where the resistive part of Z_A for a short dipole is negligible in comparison with $I/\omega C$. Also, the frequency-domain response of $i\omega CE$ corresponds to the time-domain response of $HC\dot{E}$.

Short-Circuited Loop

$$I_{SCL} = \frac{V_{oc}}{Z_L} \simeq \frac{i\omega\mu AH}{i\omega L} = \frac{\mu AH}{L} \tag{8.5}$$

where,

Z_L = loop impedance

L = inductance of loop

I_{SCL} = short-circuit loop current

The expression I_{SCL} for the short-circuit current as given in Equation 8.5 is based on the relation:

$$V_{OC} = \mu A\dot{H} = i\omega\mu AH$$

and,

$$Z_L = R_L + i\omega L \simeq i\omega L$$

since R_L is negligibly small in comparison with ωL for a short loop.

Open-circuit voltage of a dipole provides a sensitive measure of an electric field. If a good low-frequency response is desired, a very high input impedance device, such as a field-effect transitor, is required across the terminals. For high-level fields, special attention must be given at the terminals to avoid arcing. The voltage, then, must be attenuated through high-frequency attenuators or power divider circuits. Because of these difficulties, it is often more practical to use the integrated output of a short-circuited dipole for high-level fields.

While measuring a magnetic field, care must be taken to avoid resonances since measurement errors will then be introduced. Where the characteristics of the field are completely unknown, the effect of the E field on the loop can be checked by rotating the loop by 90^0 in the same plane. The response to the electric field is usually not a problem if the sensor is designed as short in wavelength, relative to the first resonance, and the major component of the magnetic field is measured in a field where $E/H = 120\,\pi$.

Current probes, as discussed earlier, can be used to measure the short-circuit current, I_{SCD} or I_{SCL}. Current probes have broadband response and low-insertion impedance, making them excellent for this application. As noted earlier, ferrite beads should be placed on output cables to reduce any current capacitively coupled to the probe from the loop, and to reduce interaction with the field.

Short dipoles and loops usually are flexible sensors in the sense that they can be oriented relatively easily to measure accurately all components of the field. All these devices must be used with systems which have frequency cutoffs below the resonance of the sensor.

Another common electric field sensor is the parallel-plate capacitor. Where the field is orthogonal to the plane of the plates, this type of sensor can be used up to frequencies whose corresponding wavelengths are much smaller than the dimensions of the plates. Reflections will occur for other directions of arrival.

8.4 Data Handling

Data handling is a major problem in EMP tests, particularly when the tests of a complex system are involved. Data taking, data reduction and analysis, data-error estimation and the formulation of data baseline are all included in the data handling process.

Before any data is taken, it is desirable to make pre-test calculation to obtain analytical estimation of test data. A complete list of measurement stations, types of measurements to be made, and the formulation of data formats should also be made before any data taking. The types and parameters of measurement, sensors and instrumentations to be used, and measurement stations or points or locations in a system where data is to be taken follow from the design of tests. As noted earlier, the design of tests is based on the test objectives.

During data taking, a visual examination of all data is necessary to determine:

- Noise at each measurement point (in the absence of signal).
- Agreement or disagreement with analytical estimates and previous data.
- Obviously inadequate sweep lengths.
- Poor calibration pulses.
- Inconsistencies among measured data at the same measurement station and for the same, or very similar, test signal, such as simulated EMP.
- Off-scale or too-small signal levels.

In general, each test should be repeated a few times to ensure data repeatability and to detect inconsistency, if any. If inconsistency is found, corrective measures should be provided, and then tests may have to be repeated. Routine raw data organization should also be followed during data taking by numbering the data in chronological sequence, annotating the data with special point of interest or pecularities observed during data taking, etc.

8.4.1 Preliminary Data Screening

After preliminary screening of the data, some calculations should be performed. These calculations should address the responses sought by the test in the predominant frequency region of interest. Commercially available plotting devices tied into calculators for automatic readout may be used, if necessary, to speed up this effort. Automated Fourier transform equipments are also available to provide rapidly spectrum of voltage or current or field, both in and out of band frequencies of interest, if simulated pulse is used for tests. Similarly, if cw test signal is used for the data, one can construct an approximate time-domain response by an automated Fourier transform equipment.

The next step in the data-handling process is data-error estimation and correction. Errors in data may be introduced by (1) test equipments, (2) uncertainties in their calibrations, (3) uncertainties in selecting time-base reference, (4) digitization errors when analog data is digitized for data reduction and subsequent processing, (5) presence of excessive noise in signal channel, and (6) numeric or computer errors during data reduction. Some of the errors will be random in character while others will be discrete. The contributions of error sources for a particular data are also not the same. Some of the contributions may be eliminated by automatic calibration mode of operation of test equipment, reversing the probe polarity and providing appropriate correction factor for the test data, etc.

The contributions of other sources of errors have to be determined by separate analysis. The contribution of an error in selecting time-base reference, for example, will be in the form of a corresponding phase-shift in the frequency-domain response. Once this error is known, its contribution in data can be readily estimated. The contribution of errors in digitizing a current waveform will introduce error in the derived Fourier transform of the response. Knowing the measure of the time-domain digitization error, one can compute the variance of the computed real and imaginary parts of the Fourier transform for any angular frequency ω.

To estimate the effect of noise on data, two types of noise should be distinguished. One is the system noise, which tends to be large, periodic and depends to a great extent on whether or not the system power is on. Although such a noise is troublesome, one can overcome the effect by timing simulated test pulses to occur in the usually available quiet-time windows. Though easier said than done, the quiet-time windows have been used in the past to eliminate the system noise as a factor in data taking. When, however, the system noise effect cannot be eliminated, the data is usually thrown out as bad data because of the large noise, and no attempt is made for error analysis and corresponding correction.

The second type of noise is thermal noise due primarily to the amplifier following the current probe. At any measuring station, its magnitude tends to be constant. The effect of such noise, therefore, is determined by signal magnitude. When making current measurements, for example, the noise effect is negligible when the signal is in the ampre range. When signals are in the low milliampre range, noise has an effect comparable to digitizing error. For still smaller signals, the noise rapidly dominates the data. For the signal range where the effect of noise on data is comparable to the digitizing error, one can replace the variance σ_i due to digitizing noise by $\sigma_i^2 + \sigma_N^2$, σ_N being the contribution toward the variance due to thermal noise, and determine the combined effect of noise and digitizing on data response or on its Fourier transform.

After appropriate correction to data has been made, the data taken for the same test-signal and same measurement station at various shots should be posted or displayed for comparison. From this baseline, useable data should be formulated and posted. Frequencies common to specific points of entry of EMP energy should be correlated at this time.

8.4.2 Data Log

Although raw data should be filed for future reference, all appropriately corrected reduced data, along with baseline data, should be posted in the data log. The data log is a continuing chronological documentary of each experimental test description, purpose and ambient conditions. It also includes a summary of comments on the response. The information thus recorded in the data log is the key to successful data interpretation and analysis.

As a minimum the data log should include:

a. test and shot number

b. date, time of day, weather

c. test object configuration description, emphasizing any specific changes, either intentional or unintentional

d. purpose of each experimental exposure to the test environment

e. general comments on the results

f. level of simulated test environment and any deviation

g. monitoring point or measurement station locations

h. checkout and calibration information

Fast retrieval of individual photographs can be established by putting them on microfilm and placing them in an automatic viewer.

From the data presented in the data log and associated comments, baseline data should be formulated. This data is necessary for overall test assessment and subsequent vulnerability analysis based on tests. At the conclusion of the test program, the total data base should be in a form suitable for direct inclusion in the final test report.

8.5 Assessment of Test Results

A comparison of test results with corresponding analytical predictions constitutes an important step for the assessment of test results. Such a comparison should be made to:

- generate confidence in test
- generate confidence in the analysis model, and
- calibrate the model

When experimental data and analytical predictions agree immediately, the conclusions are that:

a. tests had adequate control over the important variables
b. the assumed analytical model had an adequate structure and good parameter values.

When model parameters are adjusted to bring results into agreement, the model is said to be calibrated. Such a model will have a confidence level superior to that of an uncalibrated model, but it will be inferior to a model with demonstrated predictive validity. When experimental data disagrees with analytical predictions, neither data can be trusted until the disagreement is explained.

While comparing analytic predictions and experimentally-obtained test data, appropriate engineering judgment should be exercised. For example, one should note that the experimental test data should be assigned a confidence limit that reflects random test errors and data handling errors. When the confidence limit is 90 per cent, it should mean that if the same test were run many times on the same equipment, the test results would fall within the confidence level nine out of ten times. Similarly, analytic calculation should reflect all test conditions that the analytical model can accommodate. Obviously, when significant differences exist between the test condition and the model, and the effect of the differences cannot be independently estimated, the comparison loses its value.

During the comparison, one should also note that the analytical model structure is either right or wrong. There being no middle ground, no confidence level need to be assigned for the model structure. Some flexibilities, however, exist in model parameters. Depending on circumstances, analytic predictions may be given:

1. explicitly parametrically, that is, prediction based on well-defined functional relations between the prediction, and system and threat parameters

2. within a range bounded by a maximum and a minimum, or

3. with confidence limits that may result from choosing one test sample out of a larger population

Comparison between analytical predictions and test data should be done in both frequency and time domain. Knowing what part of the spectrum could be possibly in error during the test is a powerful clue to explain any difference. How significant a given difference is, however, is best judged from the time-domain comparison. The prediction which is a little low at one frequency and a little high at another may be perfectly adequate for practical purposes. It should be noted that in the final analysis, it is mostly the magnitude of voltage, current or field that is important from the viewpoint of possible vulnerability, and the exact frequency content of the response is usually not significant. A comparison of predicted and test results in minor detail is meaningless and is often distracting.

One of the principal uses of the analytical model in the final stage of vulnerability assessment is to incorporate the effects of variables not thoroughly explored during the tests. The ability of the model to track with various important variables is thus more important than absolute accuracy of either analytical predictions or test results. Accordingly, it is important that test data and analytical predictions be compared wherever possible in terms of the variations of the significant underlying variables.

Once a good correlation between test data and analytical predictions is obtained, and an understanding of EMP effects on system is established on a parametric basis, the job that remains is the vulnerability assessment of the system based on the analytical model for any specified threat or a spectrum of threats. In most cases, this amounts to the extrapolation of test data for the specified threat or threats based on the analytical model which has already been validated by tests.

8.6 References

1. Partridge, R. E., *EMP Testing Facility*, EMP Sensor and Simulation Notes, Note 1, Vol. 1, Air Force Weapons Laboratory, February 1964.

2. Baum, C. E., *Impedances and Field Distributions for Parallel Plate Transmission Line*, EMP Sensor and Simulation Notes, Note 21, Vol. 1, Air Force Weapons Laboratory, June 1966.

3. Baum, C. E., *Design of a Pulse Radiating Dipole Antenna as Related to High Frequency and Low Frequency Limits*, EMP Sensor and Simulation Notes, Note 69, Vol. 5, Air Force Weapons Laboratory, January 1969.

4. *Electromagnetic Pulse Handbook for Missiles and Aircraft in Flight, EMP Interaction 1-1*, Air Force Weapons Laboratory, September 1972.

5. *ARES Facility* (Brochure) E. G. & G.

6. *EMP Simulator Facility ALECS I*, E. G. & G., January 1969.

7. *AFWL EMP Data Reduction Handbook*, ASD 70-261, E. G. & G., November 1970.

8. *A Practical Course in EMP Data Reduction*, AL-794, E. G. & G., May 1972.

9. Baum, C. E., *EMP Simulation and Its Impact on EMP Testing*, Third Symp. and Tech. Exhibition on EMC, Rotterdam, May 1979.

10. Baum, C. E., *EMP Simulators for Various Types of Nuclear EMP Environments, An Interim Categorization*, IEEE Trans., EMC, February 1978.

11. Merewether, D. E., Prewitt, J. F., and Baum, C. E., *Characterization of Errors in Extrapolation of Data from EMP Simulator to an EMP Criterion*, Sensor and Simulation Notes, Note No. 232, Air Force Weapons Laboratory, October 1977.

12. Braun, Ch., Graf, W., Ochs, W., and Schmidt, H. U., *A Frequency Domain NEMP Simulator for Tests on Scaled Models*, Fourth Symp. and Tech. Exhibition on EMC, March 1981.

13. Giles, J. C., *Simulating the Nuclear Electromagnetic Pulse*, Military Electronics/Countermeasures, August 1977.

Index

Page No.

A

B

C

C cont'd

D

E

E cont'd

F

G

G cont'd

L

M

N

O

P

R

S

T

T cont'd

U

V

W

Y

Z